SOUND-POLITICS IN SÃO PAULO

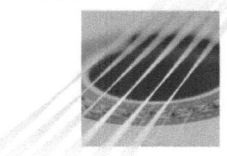

ALEJANDRO L. MADRID, SERIES EDITOR
WALTER AARON CLARK, FOUNDING SERIES EDITOR
WALTER AARON CLARK, SERIES EDITOR FOR CURRENT VOLUME

Nor-tec Rifa!
Electronic Dance Music from Tijuana to the World
Alejandro L. Madrid

From Serra to Sancho:
Music and Pageantry in the California Missions
Craig H. Russell

Colonial Counterpoint:
Music in Early Modern Manila
D. R. M. Irving

Embodying Mexico:
Tourism, Nationalism, & Performance
Ruth Hellier-Tinoco

Silent Music:
Medieval Song and the Construction of History in Eighteenth-Century Spain
Susan Boynton

Whose Spain?
Negotiating "Spanish Music" in Paris, 1908-1929
Samuel Llano

Federico Moreno Torroba:
A Musical Life in Three Acts
Walter Aaron Clark and William Craig Krause

Agustín Lara
A Cultural Biography
Andrew G. Wood

Danzón
Circum-Caribbean Dialogues in Music and Dance
Alejandro L. Madrid and Robin D. Moore

Music and Youth Culture in Latin America
Identity Construction Processes from New York to Buenos Aires
Pablo Vila

Music Criticism and Music Critics in Early Francoist Spain
Eva Moreda Rodríguez

Carmen *and the Staging of Spain*
Recasting Bizet's Opera in the Belle Epoque
Michael Christoforidis
Elizabeth Kertesz

Rites, Rights and Rhythms
A Genealogy of Musical Meaning in Colombia's Black Pacific
Michael Birenbaum Quintero

Discordant Notes
Marginality and Social Control in Madrid, 1850-1930
Samuel Llano

Sonidos Negros: On the Blackness of Flamenco
K. Meira Goldberg

Opera in the Tropics:
Music and Theater in Early Modern Brazil
Rogerio Budasz

Sound-Politics in São Paulo
Leonardo Cardoso

Sound-Politics in São Paulo

Leonardo Cardoso

UNIVERSITY PRESS

Oxford University Press is a department of the University of Oxford. It furthers
the University's objective of excellence in research, scholarship, and education
by publishing worldwide. Oxford is a registered trade mark of Oxford University
Press in the UK and certain other countries.

Published in the United States of America by Oxford University Press
198 Madison Avenue, New York, NY 10016, United States of America.

© Oxford University Press 2019

All rights reserved. No part of this publication may be reproduced, stored in
a retrieval system, or transmitted, in any form or by any means, without the
prior permission in writing of Oxford University Press, or as expressly permitted
by law, by license, or under terms agreed with the appropriate reproduction
rights organization. Inquiries concerning reproduction outside the scope of the
above should be sent to the Rights Department, Oxford University Press, at the
address above.

You must not circulate this work in any other form
and you must impose this same condition on any acquirer.

Names: Cardoso, Leonardo, 1982– author.
Title: Sound-politics in São Paulo / Leonardo Cardoso.
Description: New York, NY : Oxford University Press, [2019] |
Includes bibliographical references and index.
Identifiers: LCCN 2018037210 | ISBN 9780190660093 (hardcover : alk. paper) |
ISBN 9780190660109 (paperback : alk. paper)
Subjects: LCSH: Noise control—Government policy—Brazil—São Paulo. |
Noise—Social aspects—Brazil—São Paulo. | Noise pollution—Government policy—Brazil—São Paulo.
Classification: LCC TD893.5.B6 C37 2019 | DDC 363.74/6098161—dc23
LC record available at https://lccn.loc.gov/2018037210

This book is dedicated to my mother, Arminda Cardoso, and my father, Edmundo Cardoso.

Contents

Figures and Tables ix
Acknowledgments xi

Introduction: The Noise Multiple 1

1. *Anti-Noise Waves* 18

2. *Of Ears and Norms* 45

3. *The Echo Chamber* 75

4. *Administrative Flows* 101

5. *Legal Channels* 142

6. *The "Rowdy" Teenagers* 167

Conclusion: The Four Strata 204

REFERENCES 221
INDEX 231

Figures and Tables

FIGURES

1.1.	1932 story in the newspaper Folha de São Paulo: "The Tentacular Metropolis Is All Noise: Only the Deaf Enjoy Its Strident Life." 23
1.2.	1954 real estate ad from the newspaper Folha de São Paulo. 33
1.3.	IPT's "Urban Noise Survey Unit" (1970s). 39
1.4.	The infamous minhocão. 43
3.1 and 3.2.	Inflatable giant ear and SLM in front of the São Paulo Municipal Chamber, April 2014. 76
3.3.	Political cartoon from 2001 depicting exchange between the Evangelical councilor and the mayor. 86
4.1.	Military Police online noise complaint form 108
4.2.	PSIU's notification letter. 114
4.3.	Concrete blocks restricting access to a bar. 118
4.4.	PSIU headquarters. 121
4.5.	The São Paulo Municipal Public System online noise complaint form. 122
4.6.	PSIU form for the second offense. 125
5.1.	Area view of São Paulo's administrative center. 143
5.2.	Inside the Palace of Justice. 154
6.1.	A typical pancadão resonant machine. 179
6.2.	Map of the city of São Paulo showing the São Paulo subprefectures. 183
6.3.	Folha de São Paulo story about the first anti-pancadão operation in 2012. 194

6.4 and 6.5.	PSIU agent and other public officials. 197
6.6.	pectrograph of an excerpt of MC Guimê's "Plaque de 100." 202
7.1.	Four strata of sound-politics in São Paulo. 216

TABLES

2.1.	Five groups regularly present at the Acoustic Committee meetings for the revision of NBR 10151 and NBR 10152. 61
2.2.	NBR 10151/2000 decibel values table for daytime and nighttime according to area type. 68
6.1.	Comparison of the frequency of problems discussed at CONSEG meetings in five subprefectures between 2010 and 2012. 187

Acknowledgments

IT IS QUITE challenging to write a book about a city after living less than two years in it. Also challenging is to write a book about law, public administration, acoustics, and public security without being a specialist in any of those fields. And yet, that's precisely what I intended to do here. As I have learned from actor-network theory scholars Michel Callon, Bruno Latour, and Annemarie Mol, crossing a range of domains to follow controversies is not only feasible, but fruitful. But to survive this crossing in one piece, I had to learn to let those I encountered along the way make sense, in their own terms, of the controversies in which they were entangled. Rather than trying to compensate my lack of knowledge by critiquing the "illusive beliefs" into which my informants are miserably and stubbornly trapped, the point here is to propose a more diplomatic approach. This requires one to patiently listen, describe, and retrace all sorts of agents.

There are a few people who made this crossing possible. Colonel Wanderley Pereira, director of the São Paulo Anti-Noise Agency (PSIU) during my fieldwork in 2012–2013, generously agreed to let me visit the agency on a regular basis and accompany the inspections. Without that first-hand experience in what remains a relatively secluded and misunderstood institution, this would have been a very different book. Also working at the PSIU at the time, Debora Castelani and Lucinea Pereira were crucial figures in my study of the agency. Their patience in answering my questions over the years and their dedication to combating noise in São Paulo is nothing but admirable. Waldir de Arruda Miranda Carneiro, one of the first and only lawyers to write extensively on noise litigation in Brazil, met me on more than one occasion and, as a fine legal scholar, helped me move between legal codes, proceedings, and jurisprudence.

I was lucky to have met Yanilda Gonzalez during fieldwork. Yani, who at the time was researching the institutional reforms in the São Paulo police, helped me delineate the relations between noise and crime in São Paulo. She introduced me to the Security Council meetings and brought me along to interview key public security officials. Alexandre Barbosa Pereira, who I met at the University of São Paulo as a fellow researcher in the Laboratory of Urban Anthropology, introduced me to São Paulo's peripheral subcultures and put me in touch with the funk ostentação community.

Without the encouragement I received from my two academic mentors, I would not have the confidence to write this book. In Brazil, Elizabeth Lucas exposed me to a wide range of sociological and anthropological ideas in a stimulating academic environment. Those formative years helped me to appreciate the value of ethnographic work. In the United States, where my professional life now lies, Veit Erlmann has been an example of intellectual dynamism. Veit is one of those few scholars who can integrate ethnographic thickness and analytical breadth. He pushed me to embrace this project and saw potential in it when I could only see a cluster of loosely connected dots.

Most of my fieldwork was made possible thanks to financial support provided by the University of Texas at Austin, where I completed my PhD. Robin Moore, Sonia Seeman, Eliot Tretter, and Lorraine Leu helped to shape this project in its embryonic dissertation stage. At Texas A&M University, the Department of Performance Studies, the Glasscock Center for Humanities Research, and the College of Liberal Arts all granted generous funds for conducting additional fieldwork and covering publication costs. Donnalee Dox, my colleague and at the time head of the Department of Performance Studies, made sure I had the necessary time to focus on this book. Angie Ahlgren, Karen Bijsterveld, Veit Erlmann, Alejandro Madrid, David Novak, Marina Peterson, Matt Sakakeeny, and Barbara Weinstein offered invaluable suggestions at various stages of the manuscript. I am grateful to Antoine Hennion and his colleagues at the Centre de Sociologie de l'Innovation (the birthplace of actor-network theory) for having me in the summer of 2018 as I was finishing the book revisions. I am also grateful to Suzanne Ryan, Alejandro Madrid, and Oxford University Press for their patience and assistance throughout the process.

My cousin Cézar hosted me for more than one year in his apartment, making life in São Paulo affordable for my tight budget. My aunt Neuza, who passed away in 2013, supported me in every way she could. I will always remember her kindness. My sister was nice—and brave—enough to let me use her car (I wish I had returned it in better shape), without which this multi-sited ethnography would be less "multi" and more "sited." My parents, Edmundo and Arminda, and my siblings, Guilherme, Felipe, and Mariana, taught me early on the love for curiosity. I'm fortunate and grateful for having them in my life.

SOUND-POLITICS IN SÃO PAULO

> The group of words, police, policy, polity, politics,
> politic, political, politician is a good example of
> delicate distinctions. (Maitland 1885, 105)

INTRODUCTION

The Noise Multiple

THIS BOOK IS an ethnographic study of controversial sounds and noise control debates in Latin America's most populous city. It discusses the politics of collective living by following several threads linking sound-making practices to governance issues. Rather than discussing sound within a self-enclosed "cultural" field, I examine it as a point of entry for analyzing the state. At the same time, rather than portraying the state as a self-enclosed "apparatus" with seemingly inexhaustible homogeneous power, I describe it as a collection of unstable (and often contradictory) sectors, personnel, strategies, discourses, documents, and agencies.

My goal is to approach sound as an analytical category that allows us to access citizenship issues. As I show, environmental noise in São Paulo has been entangled in a wide range of debates, including public health, religious intolerance, crime control, urban planning, cultural rights, and economic growth. The book's guiding question can be summarized as follows: *how do sounds enter and leave the sphere of state control?* I answer this question by examining a multifaceted process I define as "sound-politics." The term refers to sounds as objects that are susceptible to state intervention through specific regulatory, disciplinary, and punishment mechanisms. Both "sound" and "politics" in "sound-politics" are nouns, with the hyphen serving as a bridge that expresses the instability that each concept inserts into the other. By denoting coexistence with the hearing body, sound *opens up* the politics of shared existence; as a matter of defining and performing the collective, politics *opens up* the acoustics of human and nonhuman associations.

The book proposes a different way of studying sound. In following noise controversies in a major metropolis through different institutional frameworks, I show the relation between the ways residents experience certain sounds and the ways certain groups (scientists, lawmakers, police officers, lawyers, judges, etc.) stabilize a definition of, and thus a "proper" way of hearing, those sounds as noise. In describing various institutional validation registers (science, politics, law, etc.) the book argues that noise is in fact multiple. I here draw on Annemarie Mol's book *The Body Multiple*. Mol argues that ontology "is not given in the order of things, but that, instead, *ontologies* are brought into being, sustained, or allowed to wither away in common, day-to-day, sociomaterial practices" (Mol 2002, 6). This is different from perspectivism. We don't have different viewpoints of a given sound, but rather different versions of it. This is also different from a type of relativism that takes each version of noise as being as valid as any other. As Mol puts it, "while realities may clash at some points, elsewhere the various performances of an object may collaborate and even depend on one another" (Mol 1999, 83). As they become controversial, sounds are constantly putting into test institutionalized validation registers (e.g., does this noise ordinance recognize this specific sound as noise?) and generating dissonance between these registers (e.g., does the predominance of this noise mean the government places religion above law?). Throughout the book, I show how homeowners, real estate and transportation groups, lawmakers, police officers, public officials, legal scholars, prosecutors, judges, and teenagers all negotiate versions of a sonic event in order to establish a reality and stabilize it as either licit or illicit. This is the book's first central claim: *noise is ontologically multiple*.

This ontological multiplicity relates to the question of why, what, how, and when one is expected to hear. For instance, as Brian Larkin argues in his study on the religious use of loudspeakers in Nigeria, auditory practices in urban centers require the constant management of attention, a "quintessentially modern phenomenon defined as an activity of exclusion" (Larkin 2014, 996). In his soundscape research, Murray R. Schafer describes modern life as being permeated by signals (foreground sounds such as sirens), soundmarks (unique community sounds such as those of waterfalls and volcanoes), and keynote sounds (constant background sounds such as the hum of an air conditioning unit). Against an urbanity permeated by loud and constant mechanical sounds and low-resolution acoustic environments (i.e., invariant and homogeneous), the author defends the preservation and documentation of meaningful, communal, natural sounds, and the more thoughtful composition of everyday soundscapes. For Schafer, the misuse of sound technology has caused the proliferation of "audioanalgesia"[1] and "schizophonia"[2] and drastically worsened our environments and the way we hear. Sonic classifications are thus embedded in notions of attention, constancy, and technological mediation. This relates, for instance, to the auditory practices that come into play when one ignores aircraft or traffic noise but considers the "leisure noise" of bars unacceptable because the former relates to work ethics and economic growth and the latter to a bohemian culture of laziness and waste.

Hearing thus requires the exercise of certain techniques of attention as forms of civility. But distinguishing "subjective" (noise as an auditory *attitude*) and "objective" (noise as legal-technoscientific *fact*) assessments of sonic events is far from straightforward. Hartmut Ising and Barbara Kruppa have suggested that sound intensity, duration, and frequency range (the three most common attributes for defining and measuring noise) cannot explain annoyance alone. Instead, "nonacoustical variables, such as situational and individual moderators, exert a considerable influence on noise processing while remaining unchanged under noise exposures" (Ising and Kruppa 2004, 8–9). Along similar lines, Jian Kang argues that a combination of acoustical and non-acoustical elements directly affect the perception of sound. These include the presence of tonal components and low frequencies, as well as the regularity and duration of quiet periods. Other aspects that influence the perception of sound as noise include fear associated with the sound, the type of activity being conducted during exposure, and how the neighborhood is perceived (Kang 2006). Karin Bijsterveld argues that European and North American legislation on noise has been entangled in a "paradox of control": lawmakers are deeply invested in measuring and controlling sounds that can easily fall into the noise-as-waste category (such as aircraft noise), but evade the more controversial community noise from neighbors, bars, and restaurants, and so on. Consequently, she argues, "Citizens have [. . .] been made responsible for dealing with the most slippery forms of noise abatement and distanced from the most tangible ones" (Bijsterveld 2008, 3–4).

Noise's ontological multiplicity also relates to issues of territoriality, always present in the act of hearing. Noise, to echo Mary Douglas (1966), is matter out of place. From preventing coppersmiths from working near intellectual activity in ancient Rome (Bijsterveld 2008, 56) to prohibiting iron-wheeled horse carriages from entering cities at night in medieval Europe (Berglund et al. 1999, iii) to regulating airport operating hours (Ashford et al. 2012), city governments have

---

[1] "The use of sound as a painkiller, a distraction to dispel distractions" (Schafer 1994, 94).
[2] "The splitting of sounds from their original contexts" (Ibid., 88).

tried to tackle noise problems by controlling space with mechanisms such as nuisance and zoning laws. But noise often disrupts these rules, permeating intricate patchworks of private and public land and raising concerns about property use. Citizens favoring individual rights demand that the state refrains from interfering with their choices of enjoying property by, for instance, engaging in the cathartic pleasures of loud music. Such a right is particularly important for nightclubs, bars, restaurants, sports organizations, and religious groups. Noise controversies as diverging expectations about the use of space can resonate with class, racial, religious, and ethnic differences.[3]

Does sound-politics include any sound? In theory, yes. But things quickly change on the ground. Why are certain sounds prohibited more quickly and forcefully than others? What justifies the authorization of loud and harmful sounds, such as those produced by airports or construction sites, next to residential neighborhoods and disenfranchised communities? What exactly is "noise" in the urban space? In *Noise: The Political Economy of Music*, Jacques Attali inverts the Marxist structure-superstructure analysis of capitalism to claim that music (the superstructure) in fact anticipates broader socioeconomic (structural) developments. For the author, music is "a herald, for change is inscribed in noise faster than it transforms society" (Attali 1985, 5), and "the noises of a society are in advance of its images and material conflicts" (Attali 1985, 11). As much as I recognize the importance of contributions by Schafer and Attali in showing new possibilities for studying noise, *Sound-Politics in São Paulo* moves in microscopic detail in opposition to such comprehensive and conjectural narratives about sound, power, and modernity. Unlike both authors, I am not interested in offering a speculative, prescriptive, or abstract conception of noise. Although this book attempts to theorize *through* sound by examining noise within power relations, it does so following from ethnographic data, thus focusing on a well-delimited geocultural milieu. By focusing on an ethnography of sounds as fluid entities entangled in a variety of social assemblages, I make the explicit effort to break away from the pre-established music-noise-silence framework on which both Attali and Schafer rely. However, as I hope it becomes clear throughout the book, more ethnographic detail does not mean less theoretical depth.

Sound-Politics in São Paulo dialogues with the explosion of interest in sound across the humanities. In the last fifteen years, sound studies has built substantial academic momentum, with the publication of edited volumes,[4] special journal

---

[3] See, for instance, Radovac (2011) on noise and class, Smith (2006) on noise and race, Weiner (2014) on noise and religion, and Boutin (2015) on noise and ethnicity.
[4] See Bull and Back (2003), Erlmann (2004b), Novak and Sakakeeny (2015), Bijsterveld and Pinch (2012), Smith (2004), and Sterne (2012).

issues,[5] and the creation of a dedicated journal.[6] Authors have shown special interest in the materiality of sound (re)production, opening lines of inquiry at the intersections of space, technology, and auditory practices. They have questioned, for instance, the premise that music and speech are the exclusive routes for understanding culture.[7] They also have challenged vision-centric narratives,[8] which often fall into sensory determinisms (e.g., hearing as more "developed" than touch but "inferior" to vision).[9] However, in the emerging literature of sound studies, analyses of the administrative and legal-scientific seizure of the acoustic register outside Europe and North America remain scarce.[10] On the one hand, as Jonathan Sterne stated recently, "the West is still the epistemic center for much work in sound studies, and a truly transnational, translational, or global sound studies will need to recover or produce a proliferating set of natures and histories to work with" (Sterne 2015, 73). On the other hand, anthropologists equipped with an "ethnographic ear"[11] have rarely entered modern institutions to examine how the state mediates listening. This book thus represents an attempt to put into use the ethnographic ear to produce a history outside but deeply connected to the West.

The book also proposes a different way of studying the state. Noise provides a unique opportunity for examining issues of governance because it permeates a wide range of institutions. Not only does noise enter the state in ambiguous ways (due to its ontological multiplicity, noted previously), but it does so—and this is crucial—at larger volumes than the state can cope with. With limited resources and often navigating within turbulent political and economic tides, public officials tend to regulate sounds that they 1) can easily specify and identify; 2) hear as imbued with negative ramifications (for health, for the economy, for political stability, for safety, for morals, etc.), especially when these public officials have the support of the citizenry; and 3) can combat under the auspices of scientific quantifiable facts. This is not an easy task, however, as all three variables are constantly shifting.

In a liberal democracy like Brazil, where the state has some level of accountability, the ubiquitous environmental noise has generated enough complaints

---

[5] See Bijsterveld and Pinch (2004), Jouili and Moors (2014), and Keeling and Kun (2011).
[6] *The Sound Studies Journal* was launched in 2015 and is co-edited by Michel Bull and Veit Erlmann.
[7] Samuels et al. (2010a, 2010b).
[8] See Erlmann (2010) and Sterne (2003).
[9] As Jonathan Sterne notes, the theological premise that "hearing leads a soul to spirit, sight leads a soul to the letter" (Sterne 2003, 16) stretches from Plato to Walter Ong.
[10] But see Ochoa Gautier (2014).
[11] This expression was used en passant by James Clifford (1986, 12) to attack Western visualism. It was later revisited by Veit Erlmann (2004b) to critique the anthropological reduction of the senses as "texts" to be read.

to saturate the state's input channels, despite the indifference of some officials. Perhaps the police officer and the judge would rather deal with more urgent offenses. Perhaps the municipal councilor and the mayor would prefer not to enrage the city's powerful Evangelical caucus by going after church noise. Perhaps, aware of the country's reliance on the car and oil industries, politicians would like to believe vehicles move as silently as they appear to do when these same officials glance at the city's traffic jams from their office windows. Perhaps the urban planner would think she can dismiss noise impact assessment in planning the city's new monorail line.[12] The political scientist, the sociologist, and the economist will claim there are more pressing issues about the future of the city worth investigating—say, voting behavior, homicide, and property values. But time and time again, residents bring noise to the equation. Unlike any other nuisance or pollution, noise is fluid, ephemeral, recurrent, and capable of entering even the most fortified upscale enclaves (Caldeira 2000). Michel Serres was right: noise is the ultimate parasite (Serres 1982, 4). This is the book's second central claim: contrary to what public officials may think, *noise is pervasive, pulsating, and parasitic. It debilitates citizen, state, and the relationship between both.*

Debates on environmental noise can help us understand how the technoscientific ideals of European urbanity continue to inform notions of civility in a global region marked by relatively high rates of violent crime, economic inequality, corruption, clientelism (i.e., patronage-dependent relations), environmental degradation, and conflicts over land ownership.[13] Compared to major urban centers in Latin America, European cities are densely populated but relatively smaller. They went through slower urbanization pace and engaged early on in debates on sustainability and environmentalism, which explains the more representative presence of acousticians and anti-noise technological infrastructure. As a result, European cities have created more robust mechanisms for tackling noise pollution. Still, the World Health Organization has alerted that one in three Europeans is annoyed during the daytime, and one in five has their sleep disturbed at night because of traffic noise (WHO 2011); although similar inquiry has not been conducted in Latin America, one can be sure the situation is worse in the region. In 2002, a European directive on the management of environmental noise required the largest cities to produce noise maps and develop noise abatement strategies.

---

[12] See "Monorail, a New Alternative to the Subway Transportation," http://www.metro.sp.gov.br/relatoriodesustentabilidade-2011/en/chapter-05/monorail-alternative.aspx.

[13] For instance, see Weyland (1998) on corruption, Helmke and Levitsky (2006) on clientelism, Liverman and Vilas (2006) on environmental degradation, and Mangin (1967) and Ortega (2004) on conflicts over land ownership.

Most of the anti-noise technology that circulates in Brazil (sound level meters and data processing software) comes from Europe. For the Brazilian acousticians I met during fieldwork, the technology that does not make it to Brazil—architectural procedures, soundproofing materials, traffic noise barriers, and so on—are reminders of the country's underdevelopment and of the ignorance and apathy of its authorities.

Also scarce are studies linking citizenship to noise in Latin America. Citizenship scholars do not discuss the problems of environmental noise as often as they do the problems of housing, crime, land ownership, voting, and public transportation. I expand on Bijsterveld's idea of the "paradox of control" and show that the *heterogeneity* of "noise" as an umbrella concept (which contains music, cars, church bells, airplanes, factories, street vendors, and more), the *complexity* of measuring it scientifically, and the *unsteadiness* of its legal encoding all make this a particularly difficult problem for the state to address. By examining how noise control plays a part in spatial arrangements, the mobilization of public institutions, and the establishment of auditory parameters, the book shows that citizenship issues involve multiple forms of belonging to, and becoming in, the Latin American city.

Citizenship scholars have noted that the gradual provision of rights in Latin America evolved differently than in the Global North. In Brazil, social rights (education, housing, fair wages, etc.) have been historically used to obfuscate political and civil rights. The centralization of power around authoritarian figures and the prominence of the executive branch in the country were sustained with a "ruling by decree" doctrine: the provision of social rights by the administration as a strategy to legitimize power and keep political opponents in check (Carvalho 2001). Roberto DaMatta argues that Brazil differs from the United States because it frames individualism in negative terms, as something that threatens the totality—a totality marked by heterogeneous and relational citizenship. For DaMatta, two powerful structures of social interaction mediate citizenship in Brazil: one based on democratic discourse that valorizes anonymity, the public sphere, individuality, and deference to the law; the other centered on personal relations, quid pro quo, friendship, and patronage, where laws and rights are modified or circumvented according to personal interests (DaMatta 1991). As James Holston notes, although this "differentiated citizenship" (where social groups use social qualifications to regulate the distribution of power) is blatant in Brazil, DaMatta's clear-cut distinction between the two structures risks conceiving of law and lawmaking as somewhat unbiased. For Holston, Brazilian law, on the contrary, "is already personalized, developing since colonial times with personalization. No special pleading is required. The individual is the seat of rights that are distributed to him or her because s/he is a certain kind of social person" (Holston 2008, 20).

The dissemination of French post-structuralism[14] and its reading of urban space as socially produced (and thereby open to political negotiation), together with intense urban migration and gentrification, stimulated a shift of focus in citizenship studies from the nation-state to the city. Stimulated by Henri Lefebvre's ideas on the right to the city, scholars have suggested that issues taking place at the city level, including public housing, transportation, and security, were crucial for understanding broader discussions about citizenship and collective rights.[15] Drawing on his study on spatial segregation and land rights in São Paulo,[16] Holston defines as "insurgent citizenship" a new paradigm of social demands that emerged in Brazil in the late 1970s, when the country was returning to representative democracy after decades of military dictatorship. In this model of citizenship, disenfranchised communities disrupt the established norm of differentiated citizenship by co-opting the democratic game and bargaining with politicians to improve their living conditions. In the 1980s and 1990s, the inclusion of disfranchised groups in decision-making spaces in Latin America through citizenship was linked to the sharp growth of nonprofit organizations and political activism at a moment of political instability and economic crisis. As Evelina Dagnino explains, in Latin America "the building of citizenship was seen at the same time as a general struggle—for the broadening of democracy—which was able to incorporate a plurality of demands, and as a set of specific struggles for substantive rights (housing, education, health, etc.) whose success would deepen democracy in society" (Dagnino 2005, 2).

Like most countries, Brazil's legislature understands that environmental noise is, for the most part, a problem more properly addressed at the city level. São Paulo's wealth, income inequality, and political history make it a fascinating field for investigating how auditory differences become articulated as rights. Left-leaning labor unions, neoliberal real-estate power brokers, conservative Evangelicals, bohemian youth, and cosmopolitan intellectuals are all well represented in São Paulo, and audibly so. In the late nineteenth century, the presence of a massive population of Italian immigrants with virtually no political rights who worked in factories under harsh conditions culminated in the country's first strikes (Khoury 1981; Biondi 2009). In the late 1970s, the massive population of factory workers in São Paulo and neighboring cities was a crucial factor in the emergence of the Workers' Party (*Partido dos Trabalhadores*, or PT), which was founded by labor union leaders, environmentalists, and social scientists (Kowarick 1985). In the

---

[14] See, for instance, Lefebvre (1991, 1996).
[15] See, for instance, Brenner et al. (2003, 2012), Harvey (1989), and Zukin (1995).
[16] See also Holston (2001).

1988 election, the first after twenty years of dictatorship, São Paulo was one of the first capitals to elect a candidate from the PT. The party has elected two presidents[17] and several mayors, including the mayor of São Paulo from 2013 to 2016, when most of my fieldwork was conducted.

In this densely packed city, it is not hard to observe how environmental noise can place liberal individualism in conflict with democratic collectivism. Whereas democratic discourse treats all noise as potentially harmful to human health, using sound level meters to obtain reliable measurements in decibels, local actors have constantly drawn attention to the more subjective ear. For them, the "signal" embedded in noise is made of "good," "necessary," "bad," and "useless" decibels. Evangelical leaders and bar owners justify their loudness by highlighting either the relevance of their holy message (as an inevitable byproduct necessary for saving souls) or the economic importance of their activity in relation to its noise. Unlike other public controversies, the concept of environmental noise emphasizes potential "polluting" activities in everyday auditory experiences while bringing together a range of sounds that exist for various reasons and affect different ears in different ways. *Sound-Politics in São Paulo* focuses on the tensions between democracy and liberty, the former based on equality and popular sovereignty and the latter on individual rights and the rule of law. That both are manifested in our current governmental frameworks does not mean they are smoothly integrated. On the contrary, I contend that the two are intrinsically irreconcilable. As Chantal Mouffe puts it, "What cannot be contestable in a liberal democracy is the idea that it is legitimate to establish limits to popular sovereignty in the name of liberty. Hence its paradoxical nature" (Mouffe 2000, 4). As it mediates individual rights, public interests, and attempts to demarcate *who* and *what* belongs to the acoustic demos, the state is constantly trying to anchor sound-politics within its regulatory domain. In so doing, it vibrates enduring cracks in the "democratic paradox" (Mouffe 2000).

Drawing on Mouffe's defense of a radical democracy that integrates irreconcilable differences in the very idea of a demos rather than emphasizing consensus, I am attentive to different ways of being political. Bruno Latour identifies five usages for the word "political." First, there is politics whenever new associations between humans and nonhumans are established, such as sound level meters or new instruments for observing the impact of noise on the inner ear. Second, politics emerge whenever an issue becomes a public problem; for example, campaigns against noise pollution. The third usage of politics comes into play "when the

---

[17] The most popular PT leader, Luiz Inácio "Lula" da Silva, is an immigrant from the Northeast state of Pernambuco and a former labor union leader in the Greater São Paulo area.

machinery of government tries to turn the problem of the public into a clearly articulated question of common good and general will" (Latour 2007, 816). An example would be lawmaking and the administrative practices designed to control noise, which rely on the ethical values of collective life.

Another possible meaning of politics refers to the process through which well-behaved citizens discuss and deliberate possible outcomes in the public sphere. In São Paulo, an example of this is the revision of the Master Plan, a piece of legislation that regulates urban growth and requires several public hearings to stimulate dialogue between state and civil society. This level of politics entails participants articulating their decisions based on "rational choice"—a notion dear to economists and political scientists. The final possible facet of politics is the stage upon which issues are black-boxed and thereby no longer seen as "political." We can include in this category the relationship between sound exposure, decibels, and hearing loss, which is virtually apolitical nowadays, with thousands of scientific studies confirming these relationships as stabilized facts.

In approaching sound-politics as an unstable series of associations mediated by the state, I draw heavily on actor-network theory (ANT). According to Latour, "ANT claims that it is possible to trace more sturdy relations and discover more revealing patterns by finding a way to register the links between unstable and shifting frames of reference rather than by trying to keep one frame stable" (Latour 2005, 24). In the last three decades, ANT has stimulated nuanced approaches to science (Latour 1987; Mol 2002), law (Latour 2010), markets (Callon 1998, 2017), and modernity (Latour 1993, 2013; Law 1986, 1991). For ANT, society is not a pre-established category with an explicative force, but a localized set of heterogeneous and temporary associations. ANT-inclined scholars retrace social assemblages by following associations between actors, which can be "*anything* that does modify a state of affairs by making a difference" (Latour 2005, 71). In so doing, these scholars describe a multifaceted network, which is never described as a pre-existing entity, but as a "*series of associations* revealed thanks to a *trial*—consisting in the surprises of the ethnographic investigation—that makes it possible to understand through what series of small *discontinuities* it is appropriate to *pass* in order to obtain a certain *continuity* of action" (Latour 2013, 33).

In that sense, whatever we understand as "noise" or "state," "sound" or "politics," entails a wide range of human and nonhuman actors, disciplinary techniques, institutional trajectories, and bureaucratic *modi operandi*, some of which I propose to detail in this book. For instance, the set of associations necessary for generating a noise ordinance includes not only the goodwill and expertise of engineers and lawmakers, but also a large number of actors such as documents, political parties, sound level meters, measurement procedures, soundproofing materials, sound amplifiers, and potential offenders—including the lovers of loud parties discussed in this book. For acousticians to become the authoritative voice in noise control controversies (i.e., the necessary path through which these controversies can be solved), actors such as traffic noise and the human ear need to be translated into decibels and public health concern. This process, which calls for the simplification of unstable scientific variables to make possible distinctions between the licit and the illicit, is often undermined by the complexity of sound perception and analysis.

To unearth noise's heterogeneous and parasitic nature, this book relies on three main sources. First, I draw on ethnographic data (semi-structured interviews and participant observation) collected during fieldwork in São Paulo in 2012, with five shorter visits between 2013 and 2017. During this period, I followed noise controversies through meeting rooms, nightclubs, and street parties; I listened to lawmakers, lawyers, property owners, and acousticians, as they attempted to reframe the city's sound-politics via laws, law enforcement, and noise measurement criteria.

The second source includes archival research in local newspapers and magazines, beginning in the early twentieth century. I took advantage of the data made available online recently and examined a large number of stories on noise in the local press. This archival research helped me to get a sense of how and when certain sounds became topics of public debate. Because the city administration is a regular topic of local press, I gained a broader understanding of how politicians have approached noise in interviews and public appearances. This also gave me a sense of what narratives were popular at a given time, and what groups were most vocal about them.

Finally, I examine an extensive network of governmental documents, including bills, laws, internal dossiers, decrees, and judicial decisions. These are the crucial governmental actors that allow us to describe the multiple channels of negotiation between the city administration, noise complainants, and noise perpetrators. Matthew S. Hull argues that these official documents have a political function that precipitates rather than simply follows the formation of networks of people and things. Like Hull, "Rather than trying to define an institution and a terrain of operations, I describe the heterogeneous relations that come into being through the use and circulation of the artifacts that mediate almost all bureaucratic activities" (Hull 2012, 21).

Chapter 1, "Anti-Noise Waves," provides a historical narrative of noise controversies in São Paulo between the 1910s and the 2010s. Drawing on the city's two most popular newspapers (*Folha de São Paulo* and *O Estado de São Paulo*), I show groups and sounds that have been central to sound-politics in São Paulo.

The chapter addresses two main issues: first, the calls from residents and groups to eliminate certain sonic practices; second, the response (if any) the city administration has provided to these calls. Are there mutual agreements between the groups and the city officials, or is this interaction permeated by disjunctures, delays, and dissonances? Retracing some of the politico-economic events in the country and the city, I argue that, since the 1910s, São Paulo has gone through six anti-noise waves, which have tended to oscillate between "behavioral" (individualized and localized, such as car honking and music listening) and "infrastructural" (collective and diffuse) sounds.

Chapter 2, "Of Ears and Norms," discusses the standardization of acoustical frameworks as crucial for the foundation of an average "public" ear, which I call "Ear 1.0." This, however, was not an easy task. The human ear is a highly specialized organ. Fragile and shielded by a thick layer of the skull, it is one of the most difficult parts of the human body to observe, which has made studying the effects of noise on human health particularly challenging. In *The Soundscape of Modernity*, Emily Thompson shows that a plethora of new techniques and instruments for controlling sound—including room reverberation analysis, noise measurement, soundproofing, and the decibel unit—helped to establish the modern soundscape in the first decades of the twentieth century as one where "sound was gradually dissociated from space until the relationship ceased to exist" (Thompson 2002, 2). Once specialists were able to build Ear 1.0, it became possible to create the sound pressure level meter (the "Ear 2.0"), an instrument capable of emulating the human ear and providing readable, comparable, and transportable data.

The second section of the chapter focuses on Brazil's two most important technical standards related to environmental noise measurement. During fieldwork, I attended dozens of meetings of acousticians, scholars, and public officials dedicated to revising these standards, which are issued by the Brazilian Association of Technical Standards (ABNT), the country's official representative in the International Organization for Standardization (ISO). Drawing on the work of Michel Callon, I argue that the presence of groups with diverse interests (the executive branch, the industrial sector, the acousticians, etc.) made the meetings true hybrid forums. During the meetings, cultural, legal, and economic rationales kept coming up despite the emphasis on scientific neutrality promoted by most participants. Debates on the best way to measure sounds oscillated between the techno-scientific and the socio-political. For example, public officials rejected requiring more accurate sound level meters in the revised technical standards because they claimed it would be too costly to purchase the devices and retrain public officials. As Brazilian acousticians have pointed out, the country's urban soundscape is the result of a lack of specialists in acoustics, a large youth population,[18] urbanization processes, construction procedures, the maintenance of a range of events throughout the year (such as *carnaval* parades and soccer games) that rely on sonic celebration, and little interest in noise pollution by environmental agencies. Again, Europe and the United States emerge as the ideal combination of professionalism, economic development, and "acoustic sustainability."

In Chapter 3, "The Echo Chamber," I consider the lawmaking process at the São Paulo Municipal Chamber. I describe some of the controversies that have informed what noise is in São Paulo and show how this problem, which entails both culturally localized practices and universalistic assumptions about public life, has surfaced in the Chamber. I describe how religious, economic, and public security issues have often mediated the regulation of certain sounds while dismissing or sheltering others. Additionally, I argue that noise debates in the Chamber illustrate how politicians deploy collective disagreements as way to advance their own agendas. Examining the trajectory of three noise-related bills in the Chamber will give us some perspective on how sound-politics intersects with groups that are directly involved with "noise" in the city. The first two bills, passed in the 1990s, address the behavioral noise of bars, restaurants, and churches. I examine how lawmakers in the Chamber who represent these groups have attempted to revise and complement both laws in order to sever any links between their activities and the state's disciplinary force. The third bill comes from the acousticians involved with ProAcústica, a nonprofit organization created in 2010. Drawing on the European Union's proactive approach to taming infrastructural noise, these acousticians have lobbied lawmakers to create environmental noise legislation centered on traffic. They managed not only to institute an annual event at the Chamber about noise but also to include a requirement for a noise map in the city's Master Plan.

Chapter 4, "Administrative Flows," focuses on São Paulo's two main administrative entities responsible for dealing with noise complaints and enforcing noise-related legislation. I start with the police, showing how officials have divergent approaches to noise (particularly its relation to crime) and solutions for it. I then discuss the PSIU (*Programa de Silêncio Urbano*), São Paulo's anti-noise agency. To understand the agency's *modus operandi* and get a broader sense of its trajectory, I conducted extensive fieldwork at the PSIU headquarters and

---

[18] According to a 2013 study conducted by the Secretary of Strategic Affairs of the Brazilian Government (SAE), the youth population (citizens between thirteen and twenty-nine years old) grew from 12.5 million in 1983 to 50 million in 2002 (26% of the total population), when it reached a pateau predicted to last until 2022 (SAE 2013).

interviewed current and previous directors. I also had the opportunity to accompany the PSIU agents on nighttime operations as they levied fines against offenders. I describe the agency's unstable condition by following the institutional shifts promoted by different city administrations. Since its creation in 1994, the PSIU has been criticized by city residents as inept, inefficient, and corrupt. A closer look at their activities will allow us to understand the tensions between the PSIU's "slow" administrative pace and the noise complainers' expectation of prompt responses to their problem.

In comparing these two law-enforcement institutions, I draw on Michel Foucault's work on governmentality to compare the configuration of authority between the police and the PSIU. Stimulated by authors such as Nikolas Rose, Peter Miller (1992) and Thomas Lenke (2013), I approach governmentality through the "humble and mundane mechanisms by which authorities seek to instantiate government" (Rose and Miller 1992, 183). By following the administrative flows that turn noise complaint into governmental punishment, I show how these flows often get held up due to tensions within and between state agencies. As the state attempts to control the environment by filtering acceptable and unacceptable sounds, these same sounds infiltrate and (re)shape the state. This double movement, embedded in the concept of sound-politics, makes both city residents and the state increasingly sensitive to sounds as objects susceptible to state intervention.

After dealing with lawmaking and law enforcement, Chapter 5, "Legal Channels," moves the discussion of sound-politics to the judicial sphere. The first section examines how a lawyer, a public prosecutor, and a judge deal with noise litigation through specific legal channels. Noise's ontological multiplicity is tangible here as well: noise-related litigation can move within the civil, public, and criminal law spheres. The second section goes through the most common strategies that plaintiffs and defendants use at the São Paulo Court of Appeals to either confirm or reverse noise-related sentences. I show that in each legal channel, the translation of noise controversy into litigation, and of litigation into effective punishment, requires the careful tying of laws and the gradual codification of reality. In considering the intersection between sound-politics and law, the chapter identifies the challenges residents face in trying to mobilize the state to use its tentacles to seize (or let go of) litigious sounds.

Chapter 6, "The 'Rowdy Teenagers,'" revisits many of the topics discussed in the previous chapters to analyze how a specific sonic practice has pushed the state to reshuffle its administrative arrangement. Whereas the previous chapters consider "environmental noise" a heterogeneous umbrella that poses particular challenges to technical standardization (Chapter 2), lawmaking (Chapter 3), administrative flows (Chapter 4), and legal channels (Chapter 5), in this chapter, we consider the peculiarities of loud music. More specifically, I discuss a type of youth street party known as *pancadão*, which became popular in São Paulo's poor suburbs in the late 2000s. To be sure, the controversy goes beyond the fact that the music played in these parties is loud. As they take place in public space, these events have become a public problem that challenges order and safety, and thereby falls under police jurisdiction. Most of the partygoers are poor and nonwhite minors, and the music played at the parties, *funk carioca,* is a style of Brazilian dance music that has historically been a target of prejudice and criminalization. Funk carioca continues to polarize attitudes towards race, youth, gender, taste, violence, and leisure in the country. It is therefore not surprising that, as it exploded in the suburbs of São Paulo and became part of the local soundscape, the pancadões affected discussions on how to better control the city's "rowdy teenagers." The term comes from James Q. Wilson and George Kelling's 1982 article known as the starting point of the "broken windows theory," which claims that the police need to address minor offenses (such as loud music) because they eventually lead to more serious crimes.

While many nearby residents call the police to complain about the loudness of the events, both police and partygoers seem aware that it is almost impossible to repress these parties with physical force. First, they are too numerous for the police to handle; in 2012, the police mapped more than three hundred parties occurring per night. Second, they rely on car speakers and therefore can easily change location. The chapter examines three main groups involved in disarticulating the *pancadão*: residents who complained to the authorities about the parties; the authorities, who responded to these complaints by organizing an anti-*pancadão* task force; and lawmakers, who started submitting specific bills to punish those involved in *pancadões*.

Tracing an actor-network requires mapping out as many institutions, disciplines, conventions, and epistemologies as is necessary. *Sound-Politics in São Paulo* unfolds through two intersecting analytical threads. The first thread relates to Latour's recent *An Inquiry into Modes of Existence: An Anthropology of the Moderns* (2013), an ambitious 500-page treatise on the diplomatic study of ontological pluralism. Following ANT's symmetrical approach to human and nonhuman agency,[19] Latour examines law, politics, religion, science, technology, and other modes of existence[20] as actor-networks imbued with specific prepositions, trajectories, translations,

---

[19] As Latour explains, "The goal will be to obtain *less* diversity in language [. . .] but *more* diversity in the beings admitted to existence [. . .]" (Latour 2013, 21).

[20] These are *not* domains or institutions. They allow domains and institutions to exist.

and trials.[21] For instance, the scientific mode enacts knowledge through extensive chains of reference, made possible thanks to inscription devices that act by making an increasingly larger number of remote beings accessible. Knowing subjects and known objects are the end results rather than the causes in this chain, with each new actor reshuffling the network as it re-articulates the points of passage necessary for constructing an objective account of the world. If science creates knowledge through straight lines of reference whose discontinuities are hidden in black boxes capable of producing "undisputable facts," the political mode operates in circles, "in a movement of *envelopment* that always has to be begun again in order to sketch the moving form of a group endowed with its own will and capable of simultaneous freedom and obedience" (Latour 2013, 134). It is the incessant performance of these circles that make issues (identity, bodies, culture, parties, the environment, etc.) matters of concern that "oblige the political to curve around it" (Latour 2013, 337).

Noise is a particularly fruitful area for discussing tensions within and between modes of existence. For Latour, "Just as, at the beginning of Fellini's *Orchestra Rehearsal*, each instrumentalist, speaking in front of the others, tells the team who have come to interview them that his or her instrument is the only one that is truly essential to the orchestra, this book will work if the reader feels that each mode being examined in turn is the best one, the most discriminating, the most important, the most rational of all" (Latour 2013, xxvi). The fact that Latour, a visual thinker who had written virtually nothing on music or sound, is now drawing on modes, tonalities, and harmonics to propose what he calls a "diplomatic" and "cosmopolitical" approach to modernity, suggests points of overlap between ANT and sound-politics. Throughout the book, I pay special attention to the scientific and technological (Chapter 2), the political (Chapter 3), and the legal (Chapter 5) modes of existence.

The other analytical thread follows the governmental arrangement by probing into São Paulo's legislative (Chapter 3), executive (Chapter 4), and judiciary (Chapter 5) branches. I consider each branch separately in order to better explain noise controversies *within* and *between* specific governmental institutions, each of which is expected to keep the others in check. The groups and institutions portrayed in the central chapters operate by performing (or attempting to perform) specific conversion procedures: sound into *scientific fact* (Chapter 2); sound

---

[21] According to Latour, "These specs, as we know, include at least four requirements: What hiatus allows us to detect the mode? What type of beings does it have? How does it differentiate truth from falsity? What particular aspect of being-as-other does it draw on to differentiate itself from the other modes?" (Latour 2013, 459).

into *instrument of political negotiation* (Chapter 3); sound into *misdemeanor or administrative infraction* (Chapter 4); and sound into *litigious matter* (Chapter 5).

Although I believe that the legislative-executive-judiciary order followed here makes the narrative less bumpy, this should *not* suggest that the book subscribes to a rigid and linear description of how the state acts. Yes, lawmakers rely on scientific data to justify the creation of new laws. Yes, the executive branch enforces the laws created by the legislature. And yes, the judiciary branch makes decisions based on existing laws and on the actions of litigious parts (including the state itself). However, as ethnographic and archival data discussed throughout this book will make evident, there is a constant influence and interference between the administrative sectors and civil groups when it comes to defining, regulating, and disciplining noise in São Paulo. Interested groups are constantly disrupting governmental arrangements, for instance, by discrediting scientific evidence, vetoing new laws, and convicting the executive for not following proper legal channels. The reason for structuring the book this way has more to do with the specificity of documents, people, ideas, and institutions examined here than with a belief the state does things according to a fixed order. Had I selected a different set of ethnographic and archival data, the narrative could have started with the judiciary or the executive branch.

In the Conclusion, I summarize the main controversies explored in the book and consider what the two analytical threads can tell us about collective living in Brazil. Can sound-politics provide a line of inquiry for examining the relations between governments and sounds? To address that question, I retrace the steps taken throughout the book to assemble sound-politics in São Paulo and put forward a more detailed analysis of how sounds have entered and left the sphere of state control. I also take a more critical stance toward the administrative practices examined here. Finally, I consider the implications of this process to our understanding of modernity and citizenship in Brazil.

1

ANTI-NOISE WAVES

SINCE THE 1870S, São Paulo has grown from a quiet colonial town to a densely-packed megalopolis. In the 1900s, it had less than 250,000 inhabitants; in the 1950s, it was one of the fastest-growing urban areas in the world; by the 1960s, it had displaced Rio de Janeiro to become Brazil's largest city. In the 1980s, it was already one of the five most densely populated cities in the world. Such an intense increase in urban density quickly exposed Paulistanos to a wide range of sounds. Some of these sounds were new, ever-present, and loud. Many residents who have had problems with these sounds called on the state to do something about it. In this chapter, we will examine controversies involving noise in São Paulo since the 1910s. To identify these controversies, we will rely on a large selection of newspaper stories.

Newspapers deal with everyday life in a city and, as a result, can consistently provide a large amount of data about any given topic. This makes it easier to identify when, why, and how certain sounds became targets of public dissatisfaction. Thus, instead of pre-determining the types of controversies worth discussing, we are going to let newspaper commentators and readers guide us. We will let their perspectives about noise in their own city lead the narrative, moving from noise to noise, complaint to complaint, and decade to decade. During this process, the ways in which civil society and the state have attempted to institute sound-politics in the city (i.e., the ways sounds became an issue of state regulation) will become distinguishable.

To establish an inventory of contentious urban sounds and examine the successive shifts in São Paulo's acoustic environment, I rely on roughly 1,510 stories that used the term ruído, or "noise,"[1] in the *Estado de São Paulo* and *Folha de São Paulo*, the city's most established newspapers. This archeological work was made possible in the 2010s when both newspapers launched online platforms with digitized editions of all issues that had been in print. Up until the 1920s, the newspapers only included the viewpoints of journalists and intellectuals. Sections such as "Varied News" (*Notícias Diversas*), "Notes and Information" (*Notas e Informações*) and "City Matters" (*Coisas da Cidade*) suggested that describing and analyzing public problems such as noise was a task for São Paulo's intelligentsia. In the 1930s, the newspapers gradually began to include their readers' personal accounts in sections such as "Letting off Steam" (*Desabafos*), "Objections and Complaints" (*Queixas e Reclamações*), "The People Complain" (*O Povo Reclama*), and "São Paulo Complains" (*São Paulo Reclama*). Rather than defining, explaining, and providing possible solutions to a given public problem, the newspapers started to assume a more intermediate position, publishing the complaints in the form of letters and giving the culprit (public and private entities) the opportunity to respond.

Like most topics that are covered by the press, noise controversies in São Paulo have appeared in waves. They gained momentum gradually, carried on by people with similar interests and solutions, reached climactic points of media exposure (with visible individuals placing considerable pressure on the state to find a solution), and faded away, displaced by other noise problems made visible by other groups with different interests and solutions.[2] In going through the noise stories in both newspapers, I identified six anti-noise waves in São Paulo between the 1910s and the 2010s.

The chapter also considers how the municipal administration framed the problems and what solutions (if any) it proposed. This will reveal some of the "translation gaps" between complainants and administrations in the back-and-forth

---

[1] "Noise" is commonly translated into Portuguese as either *ruído* or *barulho*. The Houaiss dictionary explains that *ruído* likely derives from *rugitus*, "bang" in vulgar Latin (*rugido* or *estrondo* in current Portuguese). *Barulho*, on the other hand, derives from *embarulho*, itself a variation of *embrulho*—from the verb *embrulhar*, "to embroil." Since the early twentieth century, *ruído* has been more common in the fields of science and technology, while *barulho* has a more informal usage, particularly in reference to unwanted or unknown sounds. For the search, I chose *ruído* because it is more closely related to environmental noise.

[2] In *Mechanical Sound*, Karin Bijsterveld follows some of the anti-noise waves in Europe and North America. Her analysis identifies various types of noises (community noise, aircraft noise, etc.) as problems containing a set of causes, solutions, consequences, theoretical issues, victims, culprits and owners—those responsible for framing the problem and bringing it to the public arena (Bijsterveld 2008).

between public controversies and their proposed scientific, legal, and political solutions. In examining how some local groups got their way in making noise, the chapter maps the often porous and ambiguous relations between the state and civil society.

It should be noted that this chapter will move considerably faster than the other chapters in this book. The reason for this is the amount of data that is to be covered: while the next chapters focus on the 1990s–2010s period, patiently walking from office to office and following documents until we can start envisioning actor-networks, this chapter will go through one hundred years of journalistic ink. The advantage of moving faster is that more can be introduced to the reader, specifically: 1) the individuals, groups, and institutions that have been involved in sound-politics in São Paulo; 2) the specific sounds that have been controversial; 3) the ways in which groups have mobilized actors to eliminate these sounds; and 4) the city space itself and its demographic configuration. Mapping these four components at once will provide a historical perspective and streamline the next chapters, where many of the institutions and problems that are discussed here will reappear.

First Anti-Noise Wave (1910s)

The first anti-noise wave coincided with a period of swift urbanization and densification, which was triggered primarily by a booming coffee economy, what historian Wilson Cano defines as the "coffee complex" (Cano 1981). As Cano explains, this impressive process of capital accumulation was possible thanks to land quality and availability, coffee consumption habits in Europe and North America, and a massive wave of Europeans immigrating to work as wage labor[3]—roughly one million between 1887 and 1920 (Cano 1981, 253). The coffee complex stimulated the expansion of the railroad network connecting the coast to the countryside, as well as investments in infrastructure (ports, warehouses, and public transportation). By the early twentieth century, the city had a dynamic banking system with foreign investment and emerging industrial activity. With the gradual shift from agricultural to financial, industrial, and commercial markets, as well as a rural exodus to the city during periods of crisis on coffee plantations, the city of São Paulo mushroomed.[4]

---

[3] It is important to note that the coffee boom in the São Paulo state began in the 1870s, when the vast majority of workers were black slaves.

[4] In 1890, São Paulo had 65,000 inhabitants. Three years later, the population had doubled to 130,000 inhabitants. In 1915, the city had 500,000 inhabitants, a growth of 800% in twenty-five years. In 1930, approximately 900,000 people inhabited the city.

At the turn of the twentieth century, the local press used "noise" as a sign of commercial and industrial dynamism. Paris, a model of urban life across Latin America, was the "city of noise, life, and movement."[5] Newspaper contributors listened to machine noise to attest to the growing progress of industry. They praised the sounds of shoemakers, carpenters, and smiths coming from São Paulo's Italian enclaves as a sign of the immigrants' strong work ethic. After this brief period of auditory optimism that conflated noise with economic growth, the term gradually became a sign of social problems. This acoustic shift has been discussed in the context of Global North modernity (Bijsterveld 2008, 46; Schafer 1994, 73; Thompson 2002, 115). But whereas these authors suggest that the positive and negative connotations of urban noise relate to internal tensions and shifting notions of progress in Europe and North America, in Brazil those places were seen as an ideal of economic growth and civility, a point of reference hovering above Brazil's din. For newspaper commentators, noise was the rather unpleasant evidence of the city's distance from the "developed world." In the 1910s, as São Paulo exploded into Brazil's most dynamic modern center, this auditory shift was particularly pronounced.

In 1915, a city councilor complained about city noise by describing the "shocking" experience of a foreigner visitor. This example would soon become a popular motto in the local press: *São Paulo is the noisiest city in the world!* For the foreigner, the main problem was the unceasing auditory presence of sirens, drivers and their annoying honks, coachmen shouting profanities, and prostitutes trying to get attention from passersby. These sounds, the city councilor lamented, gave the visitor "the impression of a city in the middle of *carnaval*." More concerning, he concluded, was the "lack of punishment for those who disturb public tranquility."[6] Indeed, the first anti-noise wave in the city focused on acoustic traits of "incivility." For José Manuel de Azevedo Marques, one of the leading campaigners,

> As opposed to London, Paris, Berlin, Brussels, Lucerne, Geneva, [. . .], where everything happens calmly and in relative silence, where one can have a conversation in the streets, cafes, vehicles, and offices, here it is the norm is to yell and kill. Only the insensitive, the ruffians (*pândegos*), the morons (*endurecidos*), and the idiots, for whom physiological functions prevail over moral ones, remain unshaken by this hubbub.[7]

---

[5] *O Estado de São Paulo* (August 1, 1910), 1.
[6] *O Estado de São Paulo* (March 21, 1915), 4.
[7] Quoted in *O Estado de São Paulo* (March 21, 1915), 4.

Here we have a clear-cut distinction between regressive/immoral and progressive/moral behavior in the urban space. Intellectuals such as Azevedo Marques saw themselves at the midpoint of a civilizing process by reifying two extremes: on the one hand, the "idiotic" (and mostly poor and nonwhite) residents, whose everyday noise was heard and attacked as a lower form of humanity, on the other hand, the Europeans and their silent cities, where people acted in private and public spaces with a clear sense of purpose and moral attentiveness to the surrounding collective.[8] And the issue was not confined to the public space only. A 1928 story condemned the "torturing" sounds of the radio, "which had the virtue of exacerbating the problems of its distant and archaic relative: the gramophone."[9] The main issue for the author was not the device itself, but its "vulgar" use in music stores, "filling the city with the unbearable rumble from their loudspeaker horns." One commentator explained that wide access to radio devices, made possible thanks to new credit opportunities, put the radio "within the reach of all types of people, including rude people of savage ears who enjoy music that is racket."[10]

But the biggest issue was traffic noise, which the press framed as a behavioral and, to a lesser extent, infrastructural problem. It was behavioral because drivers and streetcars used honks and bells excessively, and infrastructural thanks to the screechy cries of streetcar lines, the sudden explosions of car engines, "vehicles with iron wheels, high-pitched honks, the friction of horseshoes against the macadam, [and] the creak of the bus brakes."[11] Why couldn't São Paulo follow the example of New York City and start a broad anti-noise campaign?[12] Annoying sounds were no longer localized in one part of the city. Nor did they move in and out gradually, with the relatively slow crescendos and decrescendos of animal-powered vehicles. Traffic noise was widespread and populated the city with sudden and impulsive sounds—some (like the honk) being more intentional than others. (See Figure 1.1.)

---

[8] Azevedo Marques was not the average São Paulo resident. He served as a state deputy, federal deputy, and minister of foreign affairs. He was involved in drafting Brazil's 1916 Civil Code, and taught law at the University of São Paulo, becoming the first president of the São Paulo State Bar Association. His most important contribution to the anti-noise debate, besides regular appearances in the newspapers attacking "uncivilized" noise as a city resident and public figure, was his oft-cited article "Public Peace in Regards to Law" (Azevedo Marques 1932).

[9] "A Cidade do Ruído: Barulhos Antigos e Barulhos Novos: Automóveis, Bondes, Grammophones e Alto-Falantes," *O Estado de São Paulo* (December 23, 1928), 6.

[10] "Radio-Barulho," *Folha da Manhã* (January 23, 1938), 8.

[11] "Transito e Ruído," *Folha da Manhã* (May 12, 1938), 6.

[12] In 1929, New York City put a commission in charge of assessing urban noise by measuring the noise in several locations across the city. For a discussion of the New York Noise Abatement Commission, see Thompson (2002) and Brown et al. (1930).

FIGURE 1.1. This visual collage, published on the first page of a 1932 *Folha da Noite* issue, includes two cars, a streetcar, a gramophone, a radio, and a tuba player. The headline says: "The Tentacular Metropolis Is All Noise: Only the Deaf Enjoy Its Strident Life." The caption under the collage says: "The salad of sounds for the tympanization [sic] of the deaf in the noise city."
Credit: *Folha da Noite* (January 7, 1932), 1.

Between 1909 and 1920, virtually all noise-related ordinances and executive orders issued by the city were related to behavioral traffic noise. In a response to the shift from the carriage to the automobile, the city prohibited the use of sirens (except for fire trucks and police cars),[13] high-pitched and "melodic" car honks (only those with low-pitched and constant sounds were allowed), and motorcycles and cars with open exhausts.[14] In 1915, city councilor Luiz Fonseca submitted a

---

[13] Act 321/1909. All laws, decrees, law-decrees, etc. are from the city of São Paulo, unless otherwise stated.
[14] Act 1855/1915 and Law 2264/1920.

modified version of a bill drafted by Azevedo Marques to the Municipal Chamber. It required drivers to use acoustic warnings only once and to drive at a low speed after 9:00 p.m. to avoid disturbing residents of nearby buildings. Considered too severe or simply unfeasible by his fellow councilors, Fonseca's bill was rejected.

In the 1910s, the city started to look into other sounds as well. Act 705/1914 established hygiene and safety regulations for different types of industrial activities. Clause 31 determined that "all factories utilizing steam or electric machines that disturb the peace with noise and exude smells that bother the neighborhood" could only operate in specific areas. The act also established that hotels, restaurants, cafés, bars, confectioneries, billiard rooms, bakeries, and pharmacies could operate after statutory hours only with a special license, "which will set the closing time in accordance with the necessary peace of the population and public order."[15] In 1927, the Chamber passed a law prohibiting the use of "loudspeakers, gramophones, radiotelephonic devices, and similar instruments in music stores, shops, and commercial establishments, except [when placed] in proper chambers."[16]

The São Paulo 1929 Building Code established four main concentric zones in the city: central, urban, suburban, and rural. "The main effect of this early urban legislation," Teresa Caldeira explains, "was to establish a disjunction between a central territory for the elite (the urban perimeter), ruled by special laws, and the suburban and rural areas inhabited by the poor [left] relatively unregulated" (Caldeira 2000, 218). The first signs of spatial segregation began to appear. Working-class neighborhoods consolidated around factories and upscale residential districts in the higher flatlands. Downtown tenements formed a mosaic of dusky subdivided rooms for maximum rental occupancy. The Jardins districts,[17] praised by some newspaper contributors as a "laudable exception" to São Paulo's widespread hubbub,[18] were conceived as a garden city by private developers[19] who kept the region strictly residential, resisting apartment buildings and any threat to peace and order. In the early twentieth century, the South-West axis next to downtown (which included the Jardins) became the most valuable area of the city; insulated from poverty, pollution, and noise thanks to legislation and real estate interest.

---

[15] Act 705/1914.
[16] Law 3218/1927.
[17] The planning of Jardins ("gardens") districts (which comprise Jardim Paulista, Jardim América, Jardim Europa, and Jardim Paulistano) was inspired by Ebenezer Howard's then-popular urban theory of "garden cities." The idea was to have self-contained communities surrounded by green belts.
[18] "A Necessidade de uma Campanha Contra o Ruído," *Folha da Noite* (August 28, 1936), 1.
[19] In 1912, the City of São Paulo Improvements and Freehold Land Company, founded by European investors with experience in colonial commerce, owned 37% of the São Paulo's urban area (Rolnik 1997, 134).

Sanitation and hygiene in high-density buildings became a major concern for the state, as fears of epidemic outbreaks (yellow fever, cholera, and Spanish influenza) proved to be well-founded.[20] As Raquel Rolnik argues, "The ideological component of the sanitary movement [. . .] was one of the strongest elements constant in the urban order of Brazilian cities" (Rolnik 1997, 42). In São Paulo, the hygienists were the first group to disrupt the liberal premise of minimal state intervention in the urban space. This suggests a disjuncture of priorities between anti-noise advocates and the state. While the former saw public nuisance such as noise as a matter of civility in a limited portion of the city (the urban perimeter), the latter framed nuisance as a matter of hygiene, crucial for the preservation of the entire city. Compared to the dangerous miasmas circulating in the city, noise gained little legal and political traction.

Second Anti-Noise Wave (1930s)

The increase in complaints in the press about noise and the call for a public anti-noise campaign in the 1930s paralleled more precise descriptions of the health issues at stake. Newspaper readers now learned that loud noise in fact "destroys, little by little, the auditory organs," changes blood pressure, increases nervous and muscular tension, and even creates "gastric disorders."[21] Additionally, the permanence of noise as a public problem stimulated companies to promote their products as a partial solution to it. One starts to see ads for GM cars with "smooth and silent gears," General Electric radios with metallic valves that eliminated unnecessary noise,[22] noiseless fridges,[23] car tires that eliminated "tedious noise,"[24] car radios that "eliminate streetcar noise,"[25] Adalina Bayer pills that provided "tranquility against stressful noise,"[26] Tonophosphan injections that promised to "improve the overall health and reinforce the nervous system,"[27] and bungalow apartments located "away from the unbearable noise and annoying promiscuity of the central residential complexes."[28]

---

[20] Claudio Bertolli Filho estimates that the 1918 Spanish influenza epidemics killed 1% of the São Paulo population (Bertolli Filho 1986, 48).
[21] "Medicina e Hygiene: Os Perigos do Barulho Urbano," *O Estado de São Paulo* (March 24, 1939), 4.
[22] *O Estado de São Paulo* (September 24, 1935), 5.
[23] *O Estado de São Paulo* (February 14, 1926), 5.
[24] *O Estado de São Paulo* (November 14, 1937), 37.
[25] *O Estado de São Paulo* (May 12, 1940), 18.
[26] *O Estado de São Paulo* (February 23, 1928), 8.
[27] *O Estado de São Paulo* (September 20, 1934), 9.
[28] *O Estado de São Paulo* (December 29, 1928), 16.

In the late 1930s, the new term "decibel" began to appear in the local press.[29] From this point on, sounds were categorized according to decibels, with stories usually accompanied by a list of decibel values of everyday sounds ranging from quiet (soft conversation = 10 dB) to loud (airplane engine = 120 dB). In the 1940s, as a result, the field of acoustics began to enter the public debate. Rather than "noisy" or "quiet," a room was now described as "a set of reflections and absorptions of the sound waves, which vary according to the geometric shape of the room, the material used in its construction, and objects and even people in the room."[30] Measurability allowed the campaigners to discuss the effects of urban noise on the average person through audiometry exams.

After that point, the press began to establish a noise typology: industrial noise, school noise, aircraft noise, office noise, structure-borne noise (transmitted through solid structures), and airborne noise. This categorization increased not only the specialization of the field (bioacoustics, psychoacoustics, architectural acoustics, etc.), but the fragmentation of debates about noise, with different categories having specific effects, victims, and solutions. This increase in the visibility of sound-politics pressured the city administration to mediate and regulate a broader range of issues.[31]

The year of 1936 marked the emergence of the second anti-noise wave in São Paulo. That year, Heribaldo Siciliano appeared in several newspaper stories defending the need for a broad anti-noise campaign in the city. Siciliano, the co-founder of one of Brazil's first airlines, had been a city councilor in the 1920s and, like Azevedo Marques before him, had ties to São Paulo's politico-economic inner circle. His main platform was the *Sociedade Amigos da Cidade* (Friends of the City Society, or SAC), a nonprofit organization created in 1934 by the city's elite for the "study of problems related to the improvement of São Paulo's urban environment."[32] Under Siciliano's leadership, the SAC appeared several times in the press to propose practical solutions to urban noise, especially behavioral traffic noise. Siciliano created the Silence Commission, which asked the population to report to the Department of Transit Services any drivers who honked to get pedestrians' attention and cabs with loud honks.

---

[29] "Only six years ago," a 1939 story explained, "technicians managed to establish a norm to determine the relative intensity of sound, which will contribute to solving the problem of noise" (*Folha da Noite*, March 29, 1939, p. 3).

[30] *O Estado de São Paulo* (December 1952), 11.

[31] For instance, a 1957 report noted that the city's building code and existing laws "do not determine with the proper precision which standards to take to measure noise, smoke, smell, fire risk, and other burdens generated by the industry" (*O Estado de São Paulo*, February 10, 1957, p. 15). Another story suggested that traffic noise next to a school in the Brás district caused a higher rate of students to fail.

[32] "A Sociedade Amigos da Cidade Em Poucas Palavras," http://www.sacsp.org.br/0300-sac/

"The problem of noise [. . .] is undoubtedly a police matter!" explained Siciliano. "The police are responsible for maintaining order and have, for that reason, all the necessary resources."[33] In another story, he provides a diagram with an "indicative system of urban noise lawmaking and repression in São Paulo."[34] Among the lawmaking actors, the diagram shows the Department of Transit Service, the Civil Guard, and the Night Guard. The problem was that these law-enforcement institutions were regulated at the state rather than municipal level, generating a problem of administrative disjuncture (city versus state). Councilor and otolaryngologist Antonio Vicente de Azevedo was another leading figure of the second wave. In 1937, he suggested that the city be divided into "commercial and industrial districts, in contrast to purely residential districts, to ensure the norms of respect for the public peace."[35] His proposition was considered too strict and dismissed by the Municipal Chamber.

The premise of connecting zoning and public nuisance started timidly in São Paulo in the 1930s, as the elite realized that the urban insulation of the central districts was breaking apart. As São Paulo sprawled into an unregulated city, and as the population started to push the state to improve living conditions by de-privatizing public services and extending labor rights, the municipal government took more proactive steps to manage the city. In the 1930s, urban centers start to gain political prominence at the national level. With interstate trade wars, tensions within the military, and the 1929 financial crash, Brazil's "Old Republic" (*República Velha*), dominated by the São Paulo elite, came to an end. A new political-economic order emerged after the 1930 Revolution, a bloodless transfer of power among the political elites led by Getúlio Vargas. In tune with European trends, Vargas brought the country into an era of nationalistic authoritarianism (1937–1945).

During this period, the federal government drew on U.S. scientific management's ideas of economic efficiency and work productivity, nominating mayors invested in instituting "mechanisms that implied more planning and control over urban development" (Rolnik 1997, 166). In the 1940s, the combination of the launching of a public bus system, a housing crisis, and the low price of land in uninhabited parts of the city began to push urban growth to the peripheries. Soon, slums started to appear, with informal construction initiatives in illegally subdivided and commercialized allotments next to transportation hubs. As Nabil Bonduki puts it, "to face and conquer the periphery became an everyday task for hundreds of thousands of

---

[33] "A Necessidade de uma Campanha contra o Ruído," *Folha da Noite* (August 28, 1936), 1.
[34] "O Problema do Ruido em S. Paulo," *Folha da Noite* (December 30, 1936), 1.
[35] "O Problema do Ruido," *O Estado de São Paulo* (December 19, 1937), 11.

workers, who silently built a much larger city than the official São Paulo (Bonduki 1998, 294).

The increase in complaints against traffic noise relates to the swift increase in the number of cars and buses in the mid-1920s, partly explained by the U.S. carmakers' expansionist agenda in Latin America.[36] This national auto industry played into ideas about the link between consumerism and citizenship, as owning a car became a key component of Brazilian identity. As Joel Wolfe explains, "Broad auto ownership and two-car families continued to be more an aspiration than a fact of life in São Paulo, but those ideas reveal the power of both exotic technologies and consumer goods to shape culture" (Wolfe 2010, 42).

With car-based urbanization, the role of the city and its inhabitants as the new protagonists of economic progress hindered the creation and implementation of anti-traffic noise laws in São Paulo. Also important was the state's interventionist approach in favor of the automobile as a model of infrastructural expansion.[37] The anti-noise campaign, so popular in other metropoles, had limited success in São Paulo because it threatened the growing "traffic complex" (to borrow Cano's terminology), which included manufacturing (of cars and car parts), metallurgy, petrochemicals, and the construction and insurance businesses. Since the 1930s, such a traffic complex has provided unique growth opportunities for a cluster of industries centered in the city, and for engineer-technicians to administer this growth. By the 1950s, one-third of the city's infrastructure projects was dedicated to expressways, viaducts, tunnels, and cloverleaf interchanges.

Even if the city's middle and upper classes could move away from noise by living in strictly residential districts such as the Jardins, they still had to deal with it in their offices downtown. In 1952, city councilor Paulo Vieira unsuccessfully asked the Chamber to request the city (through the Department of Transit Services) to prohibit honking downtown, as "within this [central] perimeter there are doctors', lawyers', and engineers' offices, and the majority of government and bank branches; therefore, it's where people conduct intellectual work more intensely."[38]

---

[36] In 1925, General Motors started operations in Brazil. In 1927, it began constructing a plant in the metropolitan region of São Paulo. By 1928 it had produced 50,000 vehicles in the country. In 1928, Henry Ford successfully negotiated with the Brazilian government to build an industrial town in the Amazon rainforest to secure a source of rubber (the project was abandoned in 1934). In 1930, General Motors Acceptance Corporation (GMAC), established to finance the sale of GM cars and trucks, was launched in South America. For a discussion of automobility and modernity in Brazil, see Wolfe (2010).

[37] The Avenue Plan (*Plano de Avenidas*), conceived in the 1920s by urbanist Prestes Maia, proposed an concentric network of avenues departing from the center. As Rolnik explains, this conception of the city "was opposed to any physical obstacle to urban growth or any a priori definition of a limit for the growth of the city. This position was fully compatible with the need to spread a city considered dense and explosive" (Rolnik 1997, 161). Maia implemented parts of this plan as São Paulo mayor between 1938 and 1945.

[38] Bill 4194/1952.

Vieira's request was dismissed; then and now, traffic has proven to be anti-noise campaigners' most enduring adversary.

Third Anti-Noise Wave (1950s)

In 1951, Eucatex was founded in São Paulo. The company used eucalyptus to produce tiles, lining, and insulation materials (glass wool and wood fiber), in addition to providing soundproofing solutions for offices and factories. It advertised its products heavily in both newspapers, claiming they could eliminate 93% of the reverberation in a room and increase work efficiency by 53%. This period marks the beginning of a third anti-noise wave, which becomes discernible with the foundation of the Brazilian Acoustics Institute (*Instituto Brasileiro de Acústica*, or IBA) in 1956 in São Paulo. The IBA sponsored a series of events such as the 1956 Studies in Acoustics Week (organized together with Eucatex and the municipal administration), the 1957 symposium on industrial noise, and the 1960 "Silent Week" campaign. For the 1956 event, the IBA took an audiometer to downtown São Paulo to test the hearing of volunteering passersby, estimating that "40% of those examined were revealed to have serious hearing defects!"[39] The Silent Week included the "S-Day," which invited residents to give São Paulo twenty-four hours of silence by not "yelling in the street, not turning your TV or radio device up too loud, and not using whistles or megaphones."[40]

While the first two waves were led by people with lawmaking experience and political alliances, the third wave was represented by a group of engineers interested in what they saw as measurable, neutral, scientific facts. Here we have an important shift from noise as a behavioral/moral issue to noise as a comprehensive physical reality with quantifiable effects on human health. For third-wave campaigners, the behavioral noises of car honks, loudspeakers, exhaust pipes, portable radios, and street vendors (a focus of the first two waves) were sporadic and mostly irrelevant. The real problem was infrastructural: the "inevitable" noises of traffic and industries that were "inherent to life in big cities [and] whose effects can only be corrected with urban planning."[41] For the third wavers, it was necessary to spark an "acoustic mentality" in São Paulo, a city with "excessively

---

[39] "Decibeis," *O Estado de São Paulo* (June 6, 1958), 9.
[40] "Hoje o 'Dia-S': Tentativa para dar a São Paulo 24 Horas de Sossego," *Folha de São Paulo* (October 28, 1960), 5.
[41] "O Ruído na Cidade," *O Estado de São Paulo* (October 2, 1960), 6th Caderno.

high" noise levels "in comparison to European cities with similar populations and industrial capacities."[42]

The two first anti-noise waves tried to insulate the elite from uncivilized behavior by regulating outside noise. Third-wave campaigners attempted to invert this equation by soundproofing their offices with Eucatex materials rather than waiting for the state to discipline the streets. Moreover, whereas the first anti-noise waves were led by an elite with strong ties to the Municipal Chamber, the third anti-noise wave was centered around a group of politically neutral experts with a direct economic interest in the controversy. Interestingly, while the first two anti-noise waves resonated little within the legislature, the same was not true about the third wave.

In 1953, his first year as a mayor, Jânio Quadros submitted Bill 335 to the Municipal Chamber, proposing the control of "Urban Noises; the Location and Operation of Obnoxious, Harmful, or Dangerous Industries." In 1955, Quadros's successor, Juvenal Lino de Mattos, signed the bill into law. The document was five pages long, divided into two parts. The first part, about urban noises and public peace, seem to address the concerns of the first two anti-noise waves in that it focused on behavioral noise, including honks, bells, rattles, bugles, amplified advertisements, firecrackers, and factory whistles. It prohibited disturbing residents with "sounds considered excessive, at the discretion of the municipal authorities."[43] The second part is more aligned with the third anti-noise wave campaigners. It provided four factory categories (dangerous, obnoxious, common, and small), and determined the operating hours of each type across four zones—strictly residential, predominantly residential, mixed-use, and industrial.

In 1958, with input from the IBA, mayor Adhemar Pereira de Barros issued a decree regulating the law.[44] Barros made some crucial changes to the original 1955 document. Rather than leaving punishable sounds at the discretion of the authorities, the regulation established sound pressure levels "determined in accordance with the American Standard Association, measured in decibels with the sound level meter standardized by that association."[45] Maximum noise limits were for the first time stipulated in a piece of legislation: 85 dB for vehicles (measured at a seven-meter distance); 45 dB (nighttime) and 55 dB (daytime)[46] for loudspeakers, radios, nightclubs, restaurants, and so on; 45 dB (nighttime) and 60 dB (daytime)

---

[42] "Entidades Farão Campanha do Silêncio: há Excesso de Ruídos em São Paulo," *O Estado de São Paulo* (April 8, 1960), 11.
[43] Law 4805/1955.
[44] Decree 3962/1958.
[45] Ibid.
[46] Daytime: 7:00 a.m.–7:00 p.m.; Nighttime: 7:00 p.m.–7:00 a.m.

in strictly residential zones; 55 dB (nighttime) and 70 dB (daytime) in predominantly residential and central zones; 65 dB (nighttime) and 80 dB (daytime) in mixed-use zones; and 65 dB and 85 dB in industrial zones. The document, which expanded the original document from five to twenty-eight pages, listed every single district in the city according to these four zones.

Three important aspects should be noted regarding the 1955 law and its 1958 regulation. First, the law operated across sounds, dealing with both behavioral and infrastructural problems. This approach was important in that it recognized *any* noise as a potential problem. Because of that, it posed a new administrative challenge: what state agency should enforce such a broad law? Second, the law considered *both* time and space as indispensable parameters for regulation: sounds became licit or illicit depending on the time of day and location in which they occurred. To establish the activities allowed in each part of the city, the administration followed pre-existing urbanization trends; in bypassing a more careful analysis of noise behavior across the city, it seems the local government was more invested in protecting upscale neighborhoods from noisy commerce than in solving the problem of urban noise in the city. In any case, by informing the administration how to organize the city into chunks of time/space/activity, the noise ordinance served as the blueprint for the city's first incipient zoning law (more on that below).

Third, as the first regulation to bring measurability to the problem, it established that disputes over noise would now be mediated by a techno-scientific assemblage based on accurate and reliable *facts*. By measuring sounds, the city could now regulate areas according to certain activities, and activities according to their sounds. Despite these advances in noise control, many anti-noise campaigners claimed the law was dead on arrival—it was *letra morta*, or a "dead letter." In 1959, the city had only one sound level meter and one official trained to conduct the inspections. There was also the city's underwhelming enforcement of the law: a 1960 story reported that 600 legal cases related to this noise ordinance were waiting in administrative limbo.

Why did the city administration decide to address noise (if not in fact at least in theory) during the third wave? Did the mayors suddenly realize that city noise was a legitimate public problem? I believe the motivation here was political pragmatism. Unlike the previous two decades, when mayors and governors were nominated by a centralized federal government, in this post-Vargas period the country was transitioning to a direct elections model. From Jânio Quadros (1953–1955) to José Vicente Faria Lima (1965–1969), municipal politicians had to build an electoral platform by addressing pressing issues. Their strategies included not only appealing to real estate investors and promising basic services to the city's middle class, but also ad hoc negotiations with a growing population of marginalized city-dwellers. Jânio Quadros gained popular support by regulating and granting amnesty to illicit allotments in the suburbs, an early example of insurgent citizenship negotiations in the city.[47] Less technocratic than officials from the second and fourth anti-noise waves, public officials in this period played politics by detecting problems and strategically positioning themselves to be part of the solution.

With the elimination of central buildings inhabited by lower-class tenants during the first anti-noise wave and the expansion of the traffic network in the second wave, the third wave coincides with the consolidation of the industrial manufacturing sector and the home-ownership model in the city (see Figure 1.2). Officials at the Department of Urbanism (created in 1947) and the Advising Commission of the city's Master Plan (created in 1953) started to draw more heavily on U.S. urban planning studies such as the *Regional Plan of New York and its Environs* (New York City 1923) and *The Planning Function in Urban Government* (Walker 1941). Rather than merely a matter of organizing buildings and redesigning traffic routes, officials now saw city management as part of a comprehensive program that included a range of topics besides traffic circulation (Feldman 2005, 85).

Zoning was supposed to guide that program, regulating the continuous densification of the city, defining areas for industrial activity, and creating commercial hubs across the city to avoid excessive congestion of a small number of central districts (Feldman 2005, 94). Whether idealistic planning should follow pragmatic zoning or the other way around, remained open to fierce debate, as did the issue of whether planning and zoning were purely technical or intrinsically political matters. Either way, the city was increasingly being seen not only through a hygienist lens, but also as a lucrative spatial economy dominated by engineers, architects, and other civil construction actors who "do not desire political positions [. . .], do not participate in the administrative machine as employees, but who have links to it by participating in commissions [. . .] or by providing services to the public sector" (Feldman 2005, 78).

Despite the discourse of a more inclusive city, the zoning regulations of this period, issued via executive decrees rather than public hearings, continued to stimulate the South-West axis, densifying a few upscale districts and preserving the garden city districts. Zoning also helped consolidate the neighborhood-unity organization, with its business nucleus surrounded by residential buildings. In that sense, the noise ordinance was less a mechanism for directing urban development than one for stabilizing already existing land-use trends. It allowed the city to indirectly control property value speculation by organizing space according

---

[47] This would happen again in 1955, 1962, and 1968 (Feldman 2005, 108).

FIGURE 1.2. "Where is there less noise in São Paulo? The entire WEST side of São Paulo is a region of silence and tranquility." This 1952 ad (bottom right corner of the page) states that specialists measured the sound and reached an "astonishing conclusion": traffic noise makes it impossible to sleep at night, causing "nervous crises" among urbanites. "In almost all districts in São Paulo," the ad continues, "where one moves on cobblestone streets, noise is intense at night. However, in the WEST side [which includes the residential Jardins districts], silence is almost absolute, even with the circulation of delivery vehicles at night and in the morning, because traffic there takes place almost entirely on asphalt."

Credit: *Folha da Manhã* (December 15, 1954), 7.

to nuisance levels. Unlike the second anti-noise wave, which was mostly unsuccessful in getting the state to act due to a disjuncture of priorities, the third wave displayed a convergence of interests. If city officials wanted to organize the city without undermining existing real estate trends, noise (and nuisance more broadly) proved to be an effective parameter. In that sense, the 1955 noise ordinance was not just about tackling noise, but about fulfilling public technocrats' attempts to regulate land use throughout the entire city.

Fourth Anti-Noise Wave (1970s)

The fourth anti-noise wave roughly coincided with a period of military dictatorship, when a handful of general-presidents surrounded by technocrats controlled the federal, state, and municipal administrations. During this period, the federal government provided a strong stimulus to civil and public construction through the Federal Housing and Urban Planning Service (*Serviço Federal de Habitação e Urbanismo*), the Housing Financial System (*Sistema Financeiro de Habitação*), the Financing Fund for Integrated Local Development Plans (*Fundo de Financiamento de Planos de Desenvolvimento Local Integrado*), and the National Housing Bank (*Banco Nacional de Habitação*). In São Paulo, these programs provided cheap mortgages for the middle classes and operated through public-private enterprises to finance apartment buildings on the South-West axis.[48]

The early 1960s marked the beginning of the never-ending battle against Congonhas Airport. The root of the issue was a combination of unregulated urban growth around the airport, the airport's operating hours, an increase of air traffic, and nearby workshops used for turbine testing. Neighborhood associations soon began to demand solutions. "When we came to live here," explained a resident, "only the old German Junkers with three-piston engines were operating."[49] By the late 1960s, thanks to the new jetliners, the terms "noise" and "pollution" were starting to be used together. Middle-class residents moved away from the "contaminated" acoustic zone that surrounds the airport. Noise was entering the new public debate about environmentalism and urban sustainability. As Marina Peterson argues, "In being designated as pollution rather than nuisance, noise

---

[48] The working classes continued to occupy the periphery through informal self-constructed housing. Only in the 1980s did the administration begin to regulate areas for affordable multi-family housing projects in the suburbs. By then, the tale of two cities—a central, formal, and rich city surrounded by a peripheral, informal, and poor São Paulo—had been consolidated.

[49] "Barulho dos Jatos Atormenta Moradores de Congonhas," *Folha de São Paulo* (July 6, 1970), 4.

shifts from a relational notion, in which the source, and perpetrator, is known, to a dispersed, undefined, and atmospheric condition" (Peterson 2017, 70).

In the 1970s, civil organizations continued to condemn the major sources of air, water, and noise pollution in the city as the lack of urban planning. With that, a fourth anti-noise wave becomes distinguishable. This wave was led by the Institute of Technological Research (*Instituto de Pesquisas Tecnológicas*, or IPT). Founded in 1899 in close cooperation with European scientific-industrial institutes, the IPT fulfilled a crucial role in the city's expansion in the early twentieth century by supporting research and development of various construction materials. In 1960, the IPT's Civil Engineering Division founded a Micro-Environmental Technology Lab to address physical comfort related to temperature, lighting, and acoustics. By the mid-1970s, the IPT had one of the few acoustics labs for material testing in the country.

Like the previous wave, the fourth wave also concentrated on infrastructural noise. Ironically, it was the noise created by the municipal government itself that helped to galvanize this wave. One of São Paulo's most controversial infrastructural projects, the 2.2-mile elevated expressway known as the *minhocão* ("big worm") was conceived in the late 1960s as a solution to relieve the traffic in the downtown area across the East-West axis. The biggest reinforced concrete project in Latin America at the time, it was inaugurated in 1971 by Paulo Maluf, a young mayor with no previous political experience (his family was involved in civil construction and had founded Eucatex in the 1950s). The minhocão cuts through densely populated residential neighborhoods, with vehicles passing merely 16 feet from buildings. Since its inauguration, it has generated heated debates about state accountability in urban planning.

Together with the Congonhas Airport, the minhocão has long been an important and contentious noise pollution issue for the city. Merely two years after its inauguration, business owners and residents were already suing the city over its negative impact on nearby buildings. "The noise there is deafening; people can't have a conversation on the sidewalks at normal voice levels," asserted an IPT official.[50] By 1976, urban planners were suggesting that it would be better to tear down the whole thing in order to avoid the complete dilapidation of the street running under the minhocão. That same year, the city started to close the expressway at night in the summer due to low levels of traffic. In 1977, to avoid more legal problems, the city decided to close the expressway between midnight and 5:00

---

[50] *O Estado de São Paulo* (March 1, 1973), 30.

a.m.; in 1989, this was changed to 9:30 p.m.—6:30 a.m. Since the 1990s, it has closed on Sundays and holidays as well.[51]

In the late 1960s, the federal government started to require cities to approve a master plan as a requisite for receiving funding for urban projects. In 1965, to help plan the growing metropolis accordingly, the administration subdivided São Paulo into seven regional administrations. The administrators in charge of each regional administration would be appointed by the mayor and oversee the provision of basic public services in their area (cleaning, lighting, pavement maintenance, garbage disposal, etc.). A few years later, the city approved its Master Plan (Law 7688/1971). The law combined the concerns about traffic circulation from the 1930s with the interest in land use and zoning from the 1950s. As has been the procedure since then, the Master Plan required a zoning law. Drafted by the newly created General Planning Office (*Coordenadoria Geral de Planejamento*), the 1972 zoning law established eight different zones: three residential, two mixed, two industrial, and one for "special use." Historic downtown and Avenida Paulista area were given the maximum value on the Land Utilization Index (LUI),[52] whereas upper-class districts in the East and Southwest regions were defined as "densely mixed zones"—excluding, of course, the strictly residential Jardins area. The rest of the city was glossed over as an "immense and undifferentiated zone" (Campos and Somekh 2008, 8) where "almost anything, except big factories, [was] permitted" (Feldman 2005, 271).

Both the Master Plan and zoning law were debated by politicians and technocrats in the mayor's office and the municipal legislature and had virtually no input from the residents. Following the example of New York City, the 1972 law included a mechanism for the city to circumvent litigation with landowners by instituting a commission to decide cases not encompassed by the law. The

---

[51] In the 1990s, Maluf came back to the city administration (this time democratically elected), willing to reinvigorate his *pièce de résistance*. In 1994, for the first time ever, the city allowed buses to circulate on the elevated expressway. For one nearby resident, "one nuisance more, one less, makes little difference—we're all kind of deaf already" (*O Estado de São Paulo*, July 19, 1994, p. C1). In 1995, despite the IPT stating that noise levels would increase from 67 dB to 86 dB, the mayor proposed reopening the minhocão at night (*O Estado de São Paulo*, January 25, 1995, p. C5). In 1996, the city measured the noise near the minhocão to determine the viability of reopening it at night. The opposition in the Chamber reacted and passed a bill prohibiting the city from opening it at night. The city then looked at the possibility of walling the expressway to protect residents against the noise. Shortly thereafter, Maluf gave up. In 2015, the city decided to close the minhocão on Saturdays as well in order to encourage bikers, skateboarders, and pedestrians to use it for leisure. In 2016, the city changed its official name from "Costa e Silva" (the general-president whose 1968's decree gave the federal government the authority to force Congress into recess and intervene in state and municipal affairs) to "João Goulart," the last democratically elected president before the 1964 military coup.

[52] The LUI is a number that, multiplied by the area of a plot of land, indicated the total number of square meters on which construction could take place.

commission included major players in São Paulo's real estate construction sector such as the São Paulo Real Estate Buying, Selling, and Renting Companies Union (*Sindicato das Empresas de Compra e Venda, Locação e Administração de Imóveis do Estado de São Paulo*) and the São Paulo Construction Industry and Large Structures Syndicate (*Sindicato da Indústria da Construção Civil e Grandes Estruturas do Estado de São Paulo*).

During this period, the government insisted on framing public problems such as noise as a techno-scientific rather than a political matter. In 1968, amid international debates on ecology and the need for the state to tackle air, water, and soil pollution, the state of São Paulo created the Environmental Agency of the State of São Paulo (*Companhia Ambiental do Estado de São Paulo*, or CETESB). Affiliated with the Secretary of Health, the agency was in charge of controlling, inspecting, and regulating polluting activities. In the 1970s, CETESB and the IPT were (and to some extent remain) the techno-scientific arms of the state and the city of São Paulo, respectively. These agencies would take part in designing the city's new noise ordinance, which—as the city administration explained—would draw on the 1955 law but would be adapted to "today's needs."

If in the 1950s noise as nuisance paved the way to the zoning law, the opposite occurred in the 1970s: the city's first comprehensive zoning law led to a new noise ordinance. Clause 31 of the 1971 Master Plan determined that the city would establish maximum noise levels, measurement standards, and penalties, according to the location, time, and duration of the sonic event. In 1973, the mayor created the Pollution Control Supervision Board to plan, inspect, and educate on noise-related issues. In that same year, the General Planning Office started to discuss a new noise ordinance with the IBA and the IPT, which was to conform to the 1972 zoning law.

Bill 60/1974 established ten different noise limits: one for each of the seven zone types described in the zoning law[53] (plus a "special zone"), one for civil construction, and one for traffic. The document gave the city administration thirty days to issue a decree specifying the exact decibel values for each of the ten categories. It determined that sanctions against commercial venues would follow a three-step process: 1) warning, 2) fine, and 3) administrative closure. The document, also prepared with the help of the IPT and IBA, explained in detail which sound level meters were to be used and how measurements were to proceed. The president of the Construction Commission in the Chamber submitted an amendment to

---

[53] A—strictly residential; B—predominantly residential; C—predominantly residential of medium density and special zones; D—mixed-use of medium density; E—mixed-use of high density; F—predominantly industrial; and G—strictly industrial.

include a councilor in decibel values decisions for each category. A few weeks later, the mayor signed the bill into law[54] but vetoed the idea of including the legislature in decibel values decisions. Shortly thereafter, he issued a decree regulating the law.

Specialists from the IBA, who had helped draft the bill, complained that the final document failed to properly tackle traffic noise.[55] *Folha de São Paulo* claimed that "The civil construction sector was who most benefited from the changes in the document."[56] Indeed, in all zones construction noise for outdoor buildings was allowed to go up to 90 dB. While the original bill determined that private and public construction work outdoors could occur between 7:00 a.m. and 4:00 p.m. only, the approved document allowed certain activities to continue until 7:00 p.m. Offenders would have twenty-four hours to fix the infraction, followed by a fine equivalent to ten minimum wages per day. After the tenth fine, the construction would be shut down—alternatively, construction companies could simply pay the fine and continue working for ten days, making as much noise as they wanted.

Again, the city administration was heavily criticized for passing a dead law. By 1974, the city had only six sound level meters. Although it initially hired the IPT to conduct noise measurements (see Figure 1.3) and train new inspectors (120 inspectors and 60 engineers between 1974 and 1981),[57] by the early 1980s it seemed that the state had abandoned civil society and left it to its noise. The partnership with the IPT was discontinued, traffic and construction noise were essentially unregulated, and the inspectors the IPT had trained were moved to other sectors. "The Urban Noise Control Superintendence [. . .] [was] relegated to a tiny, empty, and quiet room of roughly eight square meters at the City Hall," reported *Folha de São Paulo* in 1983.[58] Between 1979 and 1982, the city issued just 262 fines, an extremely low number for a city of more than 8 million people. Only in the 1990s, with an international focus on environmental issues and a city administration more attentive to the average working Paulistano, would the city readdress its noise legislation.

In the 1980s, construction noise became a frequent topic in two different ways. First, residents accused construction companies of generating too much noise, to which companies responded that construction represented urban progress, and that progress could not happen without some noise. Another common excuse was

---

[54] Law 8106/1974.
[55] "Técnicos Contra o Projeto," *O Estado de São Paulo* (September 8, 1973), 1.
[56] "Lei do Silêncio Muda Nome e Espera a Regulamentação," *Folha de São Paulo* (August 13, 1974), 8.
[57] "Barulho em SP é Maior nas Marginais," *O Estado de São Paulo* (May 3, 1990), 1.
[58] "Sem Controle a Emissão de Ruídos em SP," *Folha de São Paulo* (February 20, 1983), 13.

FIGURE 1.3. IPT's "Urban Noise Survey Unit," with a microphone mounted on the ceiling of the vehicle. After the 1974 noise ordinance passed into law, the city administration hired the IPT to assess environmental noise in the city. Using this mobile lab, the IPT measured roughly 130 locations for twenty-four hours each. As Peter Barry recalls, "We equipped the microbus with a desk, a bed, and a bathroom to make the regular measurements nonstop" (personal communication, January 2017).
Credit: IPT.

the exceptionality of a given noisy event: they argued that the noise would disappear after the first construction stage, which required using a loud pile driver. Additionally, both the private and public sectors explained that they often had to work at night to avoid clogging the streets during the workday. Here again, we see the sway of the traffic complex in making people and things move *around* it to avoid disturbing its (already congested) channels.

Specialists argued that the noise problem was worse in São Paulo than in Europe and North America because of a long tradition of questionable architectural practices such as poor room organization (e.g., putting main bedrooms next to plumbing systems or elevators), extensive use of concrete (which amplifies structure-borne noise), and acoustic leaks in badly designed windows and doors. "What always seems to be of interest is security," explained Ualfrido Del Carlo, one of the founders of the acoustics lab at the IPT, referring to Brazil's obsession with thick brick walls.⁵⁹ The other way construction became a recurrent theme relates to the growth of the soundproofing market. Following Eucatex's pioneering work in the 1950s, and drawing on the research conducted at IPT and other labs, a range of acoustic treatment products became available in the 1980s. Architects suggested that homeowners stay away from noise propagators such as granite, marble, and mirrors, and instead coat walls with felt, velvet, and flannel, add a layer of plaster to the ceiling, and replace standard windows with double-layered glass ones.⁶⁰

During the fourth wave, the Brazilian National Standards Association (*Associação Brasileira de Normas Técnicas*, or ABNT), which, until then, had been a timid presence in noise debates, published two technical standards. Created in 1940 with the support of the IPT (which was concerned with the lack of national norms about construction materials), the ABNT is a private nonprofit organization officially in charge of issuing technical standards in the country. The two technical standards (to be examined in more detail in the next chapter) reinforced the idea of urban noise as both a private *and* a public problem. In the private sphere, the standard NBR 10152 (1987) established noise levels for acoustic comfort in accordance with the main purpose of the building and the rooms in it. On the public side, NBR 10151 (1987) addressed the "noise assessment of inhabited areas for the acoustic comfort of the community," including suggested decibel values for different zones (residential, industrial, etc.).

Fifth and Sixth Anti-Noise Waves (1990s and 2010s)

The fifth and sixth waves are a topic for the next chapters, so I will briefly delineate some of their elements. The fifth anti-noise wave shifted the debate from infrastructural back to behavioral issues. In 1993, *O Estado de São Paulo* stated that São Paulo was the second noisiest city in the world, with 70% of Paulistanos exposed to excessive noise every day.⁶¹ In story after story, one can read about residents complaining about loud bars, restaurants, and nightclubs. The problem was not only that venues had poor soundproofing (if any), but also that they created traffic congestion, filth, and disorder in the neighborhood. "Is the nightclub allowed to remain open after 10:00 p.m. until 5:00 a.m.? It starts on Wednesday and lasts until Sunday," complained a desperate resident in 1992. Residents also accused Evangelical churches of sonic excess with loudspeakers, shouting, and chanting.

---

⁵⁹ "Estudo da FAU Indica que Classe Média de São Paulo Mora sem Conforto," *Folha de São Paulo* (October 19, 1986), 24.
⁶⁰ "Como Vencer o Barulho," *Folha de São Paulo* (May 31, 1987), B-12.
⁶¹ "70% dos Paulistanos Sofrem com Barulho," *O Estado de São Paulo* (August 28, 1993), 6.

"I'm starting to fear for the safety of the churchgoers," vented a complainer in 1999, "because desperate apartment-dwellers are throwing objects at the church from their windows to restrain the churchgoers' [sonic] energy."[62] Bars and churches would be the main targets of the two noise ordinances passed in the 1990s, which I discuss in Chapter 3.

The sixth and final wave began in the early 2010s when another nonprofit organization took the lead and moved the debate to infrastructural noise once again. Created in 2010, ProAcústica is a market-oriented nonprofit organization that brings together companies and specialists "willing to leverage the development of applied acoustics in Brazil."[63] Like the IBA in the 1950s and the IPT in the 1970s, ProAcústica lobbies for the creation of noise legislation and the revision of technical standards. But rather than focusing on products for property owners, the organization has tried to push construction companies to change their modus operandi by including soundproofing materials and techniques in their new buildings. Additionally, ProAcústica has successfully convinced lawmakers to create an annual debate on noise and to create a law for mapping traffic noise in the city.

Still, behavioral noise continued to bother residents. Complaints about loud music coming from street parties increased across the city in the late 2000s, particularly in poor suburbs. Like the debates against bars in the 1990s, residents framed their complaints by drawing a clear contrast between the city of work (with which São Paulo has been associated since the turn of the twentieth century) and the city of leisure; the city that builds things and brings economic progress, and the city that wastes things in hedonistic behavior. As I discuss in Chapter 6, this controversy resonates with debates about criminality, police activity, youth behavior, and urban segregation.

WHY SO NOISY?

By looking at complaints in two local newspapers, we have detected not only the types of sounds that have bothered residents, but also the institutions, technologies, documents, and groups that have mediated debates about those sounds. Although one debate would not necessarily fade away before others entered the scene, the stories examined here suggest that the city has gone through a series of anti-noise waves with new ones emerging roughly every two decades: in the 1910s (with Azevedo Marques), in the 1930s (with Siciliano and the SAC), in the

---

[62] *O Estado de São Paulo* (February 3, 1999), Z2.
[63] Quem somos, visão, missão e valores, http://www.proacustica.org.br/sobre-a-associacao/quem-somos-visao-missao-e-valores.html

1950s (with Eucatex and IBA), in the 1970s (with IPT and ABNT), in the 1990s (with the city administration itself), and in the 2010s (with ProAcústica). The repertoire of controversies gathered in these waves allow us to tentatively address two fundamental questions: *Why does urban noise persist? Why should the state attend to it?*

Regarding the first question, the stories analyzed here suggest four recurrent tropes. First, noise persists because of a general *lack of human civility*. The poor, the "ruffians," the "morons," the "idiots," the immigrants, the poor, the nonwhite, and the youth are simply not civilized enough to understand the tacit (let alone the legal) rules of urban living. This premise is particularly evident in the first two waves, when the new mechanical sounds of car honking, radio devices, and gramophones (in the streets, stores, and households) introduced new possibilities for injecting personalized sonic markers into the public space. To the chagrin of the anti-noise campaigners invested in making São Paulo Latin America's most modern metropolis, such a seizure of the acoustic public sphere pushed the city toward the "barbaric" rather than "cultured" end of the spectrum.

The second cause, evident in all six waves, is *governmental leniency*. Residents continuously fail to behave in a civilized manner partly because the state continuously fails to discipline them, creating a mutually reinforcing circle. The notion of impunity is apparent in different forms, including nonexistent or badly designed legislation, inefficient law enforcement, a corrupted judiciary, and a lack of investment in staff and technology. We find this reasoning back in 1934: "In poorly policed cities, the sacred rest of one's neighbor is not respected"[64]; in 1955: "The lack of law-enforcement generates abuses"[65]; in 1976: the city is unable to handle noise complaints "due to a lack of technicians [and] the necessary amount of sound-measuring devices"[66]; in 1999: "Are public services (authorized by the city) to be completed with a jackhammer working in the middle of the night, or can we treat the taxpayer with a minimum of seriousness and respect?"[67]; and in 2008: "It takes [the city administration] ten minutes to answer the phone; they ask several questions, say they're going to measure [the noise level], and nothing happens after that."[68]

The third cause relates to *infrastructure deficiencies*. The noise of streetcar rails installed over an irregular paving stone, faulty asphalt, bumpy streets, ungreased subway rails, traffic jams, inadequate bus and airplane routes, and ineffective urban planning, all seem to contribute to São Paulo's cacophony. Infrastructural

---

[64] "O Ruido nas Cidades," *Folha da Manhã* (September 20, 1934), 5.
[65] *O Estado de São Paulo* (January 18, 1955), 3.
[66] "Ruído é Ainda Incontrolável," *O Estado de São Paulo* (November 12, 1976), 23.
[67] *O Estado de São Paulo* (May 21, 1999), Z2.
[68] *O Estado de São Paulo* (January 11, 2008), C2.

FIGURE 1.4. The infamous minhocão.
Credit: Google Street View. https://www.google.com/maps/@-23.5365395,-46.6505254,3a,75y,306.23h,93.04t/data=!3m6!1e1!3m4!1suM59dz1DBNBwiVAHLTWJrA!2e0!7i13312!8i6656

problems resonate particularly with the second (1950s), fourth (1970s), and sixth (2010s) anti-noise waves, when sound specialists called for the revamp of city spaces through the control of major sources of noise. Campaigns against Congonhas Airport (starting in the 1950s) and the minhocão (starting in the 1970s) become quests for better assessment and regulation of urban infrastructure (see Figure 1.4). Unlike the first two explanations, which were more accessible to the average citizen, discussions about infrastructure deficiencies is a territory in which sound specialists reign supreme. It gives them leverage to steer public officials' interest toward technical solutions that require a specialized workforce.

The fourth and final explanation, closely related to infrastructural deficiencies, is *technological obsolescence*. For most of the twentieth century, this explanation resonates with the belief on the "technological fix," or the premise that "modern science, engineering, and machines would resolve social problems that societies would not otherwise be able to rectify" (Wolfe 2010, 9). Older technology is the source of noise problems and the solution requires the implementation of newer technology. For example, in 1975, the problem of aircraft noise in the Congonhas Airport region was explained as a lack of "more modern equipment."[69] Like infrastructural deficiencies, this fourth cause falls under the sound specialists' domain. However, unlike the former cause, which can easily intersect with other public problems (air pollution, traffic congestion, urban planning, etc.), noise

---

[69] "A Paz Noturna está Voltando a Congonhas," *Folha de São Paulo* (February 19, 1975), 10.

control technology is considered costly because it is specific, making it harder for specialists to appeal to public officials.

There is little doubt that, in these four explanations, as expressed in the six anti-noise waves analyzed here, Europe and the United States remain the points of reference. Residents, specialists, and politicians marvel at these developed nations ranging from optimism (they've got it, so can we), to pessimism (we're never going to get it), and to realism (here's what we can do). This type of reverence for all things north of the equator came to the fore repeatedly. Paulistanos constantly allude to the United States, Germany, France, and England when complaining about drivers' honking habits or nightlife hubbub, calling for new laws (or attacking old ones), demanding better urban planning, praising quieter products and soundproofing procedures, and urging social change.

The second question—why the government should get involved—can be explained by the fact that noise touches on three paradigms of social order: peace, rest, and health. The citizen has the right to peace, particularly on her own property. More than an issue of nuisance, peace can touch on property rights. As a powerful regulator and mediator of property rights, the city administration is held accountable whenever it disrupts—or allows others to disrupt—the city's real estate arrangement. The second paradigm has to do with an economic rationale that asserts that workers need to rest so they can perform properly. It is not surprising in São Paulo, which for decades has led the country's economy, rest and work ethics are often prioritized. Third, the state is accountable for overseeing public health, which includes noise pollution. To the extent that health specialists establish a causal relationship between environmental noise and health problems (hearing loss, heart disease, changes in the immune system), they push the state to intervene as the ultimate manager of the body politic. As we will see throughout the book, peace, rest, and health have different legal weights, partly because each relies on different degrees of measurability and causality reasoning.

As these waves moved back and forth from civil to governmental actors, they encompassed notions of "disturbance of the peace," "nuisance," and "pollution," growing into a highly heterogeneous public problem that included car honks, motorcycle exhaust pipes, gramophones, radios, factories, Congonhas airport, the minhocão, restaurants, nightclubs, pile drivers, and buildings, among other things. Now that we have a better sense of how São Paulo sound-politics has unfolded over the last one hundred years, the narrative can more properly focus on the last decades. In the next chapters, we will further zoom in on certain debates and slow down the narrative. Our goal is to follow controversies about technical standards, ordinances, fines, and judicial decisions: the four tentacles through which the government has tried to stabilize acoustic dissent in São Paulo.

2

OF EARS AND NORMS

THE PREVIOUS CHAPTER mapped a range of acoustic divergences in São Paulo since the 1910s. Looking at the use of the word *ruído* in the city's most popular newspapers, we identified six anti-noise waves, each involving specific sounds, groups, institutions, and regulations. This mapping showed that, particularly since the 1930 Revolution, noise as an urban problem tended to move back and forth between the specialists and the nonspecialists, behavioral and infrastructural issues. Wave after wave, a growing group of acoustics and public health specialists brought debates on noise control into a technical arena. Experts associated with the IBA (Brazilian Acoustics Institute), the IPT (Institute for Technological Research), and ProAcústica have tried to become the necessary points of passage for tackling urban noise by maneuvering the controversy away from the bitter politics of nonspecialists. It is now time to abandon the mapping approach and look at things closer to the ground. How precisely (and successfully) have these experts performed this maneuver? How have they managed to mediate between noise and noise control, problems and solutions?

This chapter focuses on several attempts to mobilize sound-politics via a range of techniques, devices, and national and international standards. To follow noise controversies in São Paulo we need to consider how experts have tried to stabilize two central actors: ears and norms. The most powerful argument for governmental intervention in noise problems has been evidence of its harmful effects on the human ear. By examining the ear and putting it through a series of tests, experts have generated an "average normal ear," which I call *Ear 1.0*. To a large extent, the configuration of sound-politics depends on the strength of the links between noise, Ear 1.0, and public health initiatives.

There is, however, one link missing in this chain. The second actor is the black box responsible for emulating Ear 1.0 and conveying reliable, quantifiable, and transportable information. The sound pressure level meter (SLM), or *Ear 2.0*, is crucial here because it can convert Ear 1.0 into what Bruno Latour and Steve Woolgar call an "inscription device," that is, an "item of apparatus or particular configuration of such items which can transform a material substance into a figure or diagram which is directly usable" (Latour and Woolgar 1986, 51). An inscription device requires someone (an expert) to *speak for* the device, translating and interpreting the information that shows up on the display screen. The SLM promises to bypass "subjective" and unreliable results by folding into its circuitry the thousands of ears amassed by Ear 1.0. As other authors have already traced the fascinating history of the stabilization of Ears 1.0 and 2.0 in the early twentieth century, the first section of this chapter will simply summarize some of these debates.

Building on the discussion about Ears 1.0 and 2.0, the second section focuses on the Brazilian Technical Standards Association (ABNT). In the 2010s, experts faced the daunting task of revising two technical standards for assessing environmental noise. As I show, drawing on participant observation, interviews, and minutes of meetings between 2011 and 2017, the endeavor was challenging because these revisions involved the input of groups with different interests and different understandings of what a technical standard should do. Despite the wish of many for a purely "technical" approach to noise measurement procedures, establishing the *norm* was permeated by political, administrative, and legal concerns. Drawing on Michel Callon, I argue that this heterogeneity of viewpoints made the ABNT meetings hybrid forums.

Ear 1.0 and Ear 2.0

It comes as no surprise that foundational sound studies texts focus so much on the 1870s–1930s period. Besides intensive industrialization, urbanization (often permeated by audible inter-ethnic tensions),[1] and mechanization, this is when we start to see a "paper trail left by sound-reproduction technologies" (Sterne 2003, 7). Gradually, with the global emergence of new technological entities for

---

[1] See Weiner (2014), Boutin (2015), and Bijsterveld (2008).

mastering sound, "the many different places that made up the modern soundscape began to sound alike" (Thompson 2002, 3). This is the period of consolidation for New Acoustics, a field centered on "new tools, new techniques, and a new language for describing sound" (Thompson 2002, 5).

The emergence of New Acoustics took shape thanks to the convergence of three groups. The first comes from medicine and includes mostly physiologists and otologists. The small size, internal specialization, fragility, and difficulty of access combine to make the human ear a remarkably difficult object to study. Békésy and Rosenblith (1948) summarize the gradual understanding of the ear in five foundational moments, each based on specific techniques and technology. In the first moment, up until the sixteenth century, specialists described the ear mostly based on observations of the external ear and eardrum. The second moment was the period of "gross anatomy," represented by the work of Andreas Vesalius (1514–1564), Bartolomeo Eustachio (1510–1574), and Hieronymus Fabricius (1537–1619). By shattering the temporal bone of cadavers to examine the ear, these experts were able to probe into the middle ear ossicles, the auditory nerve, the cochlea, and the semicircular canals.

The third period, which started around the seventeenth century, focused on the inner ear, now accessible thanks to the miniscule chisels developed by Italian goldsmiths. Experts such as Thomas Willis (1621–1675) and Antonio Maria Valsalva (1666–1723) established a distinction between the outer ear, middle ear, and inner ear, and identified the cochlea as the actual locus of hearing. The fourth period was linked with the work of Antonio Scarpa (1747–1832). Based on microscopic observations of the cochlea, Scarpa "followed the course of the auditory nerve from the ear to the brain with a precision unknown before his time" (Békésy and Rosenblith 1948, 735–736). Finally, in the last period identified by the authors, experts increasingly began to rely on a variety of procedures, including dental burrs, comparative experiments with living animals, and recordings of electrical impulses. The corrosion technique of Joseph Hyrtl (1811–1894) allowed for the preservation of the cavities of the ear by filling them with wax. Marchese Alfonso Corti (1822–1875) identified the sections of the cochlea (one of which bears his name) and "the dimensions of the various parts of the [cochlear] membrane" (Békésy and Rosenblith 1948, 736). Hermann Helmholtz postulated that "individual nerve fibers acted as vibrating strings, each resonating at a different frequency" (Rossing 2007, 13).[2]

The unearthing of the human ear was paralleled by an interest in the boundaries of human hearing in regard to amplitude and frequency perception.

---

[2] For an analysis of the relations between music aesthetics and theories about the ear, see Erlmann (2010).

In the early 1880s, Félix Savart used a rotating toothed wheel to determine the range of hearing frequencies. Similar studies were conducted by William Hyde Wollaston (in 1820), ésar-Mansuète Despretz (in 1846, using tuning forks), and Francis Galton (in 1876, using a brass whistle later known as the Galton whistle). Establishing the amplitude limits from the minimally audible to the painful proved to be more challenging due to the lack of devices capable of generating quantifiable intensity units. Up until the emergence of electric circuits, experts relied on various types of hammers, tubes, and tuning forks. In 1885, Arthur Hartmann designed an auditory chart, using tuning forks to measure percentages of hearing. In 1903, Max Wien also used tuning fork stimuli to develop a sensitivity curve, in one of the earliest attempts to measure the relationship between intensity and frequency thresholds (Vogel et al. 2007, 82). Although these experiments were crucial for the construction of Ear 1.0, the lack of measurability remained an obstacle for the specialists.

The second representative group in the "New Acoustics" field came from physics and centered on the study of sonic behavior. This group included well-known figures such as Pythagoras,[3] Galileo Galilei,[4] Ernst Chladni,[5] and Pierre-Simon Laplace.[6] In 1862, Helmholtz published *On Sensations of Tone*, in which he identified the presence of upper partials along with a fundamental frequency. In that same year, Rudolph Koenig observed acoustic signals using an oscillating flame and a rotating mirror. Twenty years later, Baron Rayleigh used suspended disks to determine the amplitude of a sound source. At the turn of the twentieth century, Wallace Sabine's measurements of reverberation time and the absorption coefficient of different materials helped to consolidate the field of architectural acoustics (Thompson 2002). By instituting the field of acoustics, this group established yet another chain of reference connecting sound (now an object with specific properties) and hearing.

The final and third group of New Acoustics was made up of the sound media entrepreneurs who by the 1930s had already built telecommunication conglomerates with global ambitions. Drawing on experiments from the first two groups, these explorers helped to turn sonic events into stable object-signals that could be transmitted, stored, reproduced, and amplified. As Sterne notes, this process included three main events:

---

[3] Pythagoras used the monochord to identify frequency ratio relations.
[4] In the sixteenth century, Galilei used the pendulum and vibrating bodies to establish the relationship between frequency and pitch.
[5] In the eighteenth century, Chladni's vibrating plates made the nodal lines of soundwaves visible.
[6] In the 1810s, Laplace conducted a series of investigations about the speed of sound.

(1) the emergence of audile technique as a way of abstracting some reproduced sounds (such as voices or music) as worthy of attention or "interior," and others (such as static or surface noise) as "exterior" and therefore to be treated as if they did not exist; (2) the organization of sound-reproduction technologies into whole social and technical networks; and (3) the representation of these techniques and networks as purely natural, instrumental, or transparent conduits for sound. (Sterne 2003, 25)

The 1870s was a particularly remarkable decade for sound media exploration. Trying to build a telegraphic device that could send multiple messages via tuning forks, Alexander Graham Bell started to work on the idea of transmitting sound information as a continuous rather than an intermittent current. In 1874, inspired by Koenig's manometric flame and Édouard-Léon Scott's 1857 phonoauthograph (a device that used an elastic membrane and a stylus to inscribe sound waves on glass or paper), Bell created an ear phonoauthograph by mounting human ear ossicles and an eardrum onto a wooden frame. Later in that same year, he started to conceive of the use of a diaphragm attached to an electric current in order to transmit sounds over distance.

In 1877, one year after Bell obtained his first patent for the electric telephone, Thomas Edison publicly announced his phonograph. Also drawing on Scott's phonoauthograph, the device could record and reproduce sounds engraved on a thin foil. In that same year, Edison won a patent for the microphone, which had two metal plates separated by granules of carbon. The device transduced sound pressure into electric signals by vibrating the granules, causing changes in electric resistance. Also in 1877, Ernst W. Siemens received a German patent describing what would later become the loudspeaker: a moving-coil transducer "with a circular coil of wire in a magnetic field and supported so that it could move axially" (Rossing 2007, 18). By the end of the nineteenth century, Edison had founded Edison General Electric (later the General Electric Company), and Bell had established a vast telecommunications complex, which included Bell Labs, the Western Electric Company, and the American Telephone and Telegraph Company (AT&T).[7]

The convergence of experts from medicine, physics, and telecommunication into New Acoustics was further consolidated in 1929 with the creation of the Acoustical Society of America. It is here that we will find the inauguration of Ears 1.0 and 2.0. As telephone lines started to spread, it became financially imperative to measure

---

[7] I am here less interested in reproducing a canonized narrative about inventors and inventions than in following figures whose *legal hold* over the production and distribution of these objects determined the economic power of the telecommunication conglomerates they set in motion.

signal loss over distance. In 1904, Bell Telephone Laboratories designed the MSC (Miles of Standard Cable) unit, with 1 MSC at 795.8 Hz being the slightest difference of intensity detectable by the average listener.[8] Tests suggested that "commercially acceptable speech was achievable over a connection of 46 MSC [. . .]; connections not more than 50 miles apart should be no worse than 30 MSC" (Ward 2006, 32). In 1924, the company replaced the MSC with the TU (Transmission Unit). This unit

> was a distortionless, logarithmic unit so chosen as to make use of common logarithms convenient in transmission computations. Its magnitude was very nearly the same as the loss of a mile of standard cable and, thus, existing experience learned in terms of miles of standard cable could be transferred to the new system with a minimum of difficulty. (Sullivan 1971, 2669)[9]

In 1928, the TU became the decibel (dB), with 0 dB (10^{-12} W of sound power) becoming the reference value. As the ratio between the weakest perceptible sound and the pain threshold is rather large (10^{12}), it seemed logical to convert these numbers into a logarithmic scale. The decibel is a practical device to express the difference between a reference value and a specific value to powers of 10. For instance, if you put ten identical washing machines in a room that had one machine, you would be multiplying sound pressure by 10 and increasing the sound level by 10 dB. Increasing the intensity by a factor of 100 corresponded to a 20-dB increase (a 30-dB increase would require 1,000 washing machines in the room). A change in the power ratio by a factor of 2 results roughly in a 3-dB variation. In other words, if you put twenty washing machines in a room that had ten machines, the sound level would increase by 3 dB.

If the decibel needs a reference value to operate, then what is the minimum threshold for the "average" listener? In the 1920s, AT&T continued to refine telephone communication by building Ear 1.0 through several audiometric surveys, a process Mara Mills refers to as the "ergonomopolitics of objects" (Mills 2011, 130). In 1922, Western Electric workers Robert Wegel and Edmund Fowler coined the

---

[8] As Mara Mills describes, this research on the "just noticeable difference" (JND) between sounds drew on experiments by Edward Wheeler Scripture and Carl Seashore in the 1890s. Seashore developed the first commercially successful audiometer in the United States (Mills 2011).

[9] As John Hilliard explains, "The 'TU' also eliminated the problems associated with the mile of standard cable (which had been used for twenty years) which included [frequency response] distortion due to its inherent inductance and capacitance. The 'TU' could then be independent of frequency so that it could be used to measure power ratios at any frequency. The sound power changes that could be detected by the ear corresponded to the 'mile of standard cable.' The properties of the 'TU' made it available to other fields and was not restricted to telephone circuits" (Hilliard 2006[1984], 1).

term "audiogram," showing the "variation of minimum audible sensitivity with frequency" (quoted in Mills 2011, 129). One year later, Western Electric produced the first widely-used audiometer (the Western Electric 2-A), which included eight frequencies, but lacked a calibration for amplitude levels. In 1933, Harvey Fletcher and Wilden A. Munson proposed average equal-loudness contours, based on a survey where subjects wearing headphones were asked to determine the equal loudness of pure tones in various frequencies, using a 1 kHz tone as the reference (Fletcher and Munson 1933). In retrospect, using 1 kHz as the zero-point proved not to be the best choice, as the human ear is most sensitive between 2 and 5 kHz, leading the chart to include negative decibel values. Between 1935 and 1936, the U.S. Public Health Service conducted a series of hearing surveys to establish the audiometric zero reference standard. In the 1950s, D. W. Robinson and R. S. Dadson proposed a re-determination of the equal-loudness contours (Robinson and Dadson 1956), which served as the basis for the international standard ISO/R 226 (1961). Their 1956 study proposed substantial changes to the Fletcher-Munson curves, with a discrepancy of up to 14 dB below 500 Hz. However, a 2003 revision of ISO 226, which relied on studies conducted in Japan, Germany, and Denmark with individuals between eighteen and twenty-five years old, showed that the 1933 Fletcher-Munson contours are in fact more accurate than the Robinson-Dadson curves, particularly in the higher frequencies (Suzuki and Takeshima 2004). In 2014, the Acoustic Technical Committee reviewed and confirmed the 2003 version of ISO 226.

With a (somewhat) stabilized Ear 1.0 and a workable quantification system in place, it became feasible to build Ear 2.0. For the anti-noise campaigners, who were eager to tame urban noise but still lacked an inscription device capable of bypassing the unreliable individual ear, Ear 2.0 was a game changer. In 1926, Edward E. Free conducted measurements across New York City by adjusting the pure tone of a Western Electric 3-A audiometer to equal the loudness of the measured noise (Thompson 2002, 148). In 1929, New York City's Noise Abatement Commission conducted 10,000 measurements at 138 locations (Thompson 2002, 148).[10] In a 1930 article, Free described the "acoustimeter" recently developed by C. F. Burgess Laboratories, which provided "the most flexible, accurate and reliable method of noise measurement so far available" (Free 1930, 29). The device included a condenser microphone, vacuum tubes to pre-amplify the signal, battery boxes, and a galvanometer showing the readings in decibels. It captured sounds between 60 and 10,000 Hz, and between 0 and 120 dB. Drawing on research on equal-loudness contours, the acoustimeter included an electric network of condensers

---

[10] For a discussion of the New York City Noise Abatement Commission, see Wynne (1930).

and inductances that emulated human ear responses at different frequencies. The emulation was an approximation, as it would be "impossible to produce networks which will correspond exactly to any one definite curve drawn to represent the sensitivity of the human ear" (Free 1930, 27).[11] Additional filters could be attached to the device to measure noise at specific band frequencies (20–200 Hz, for example). These early sound level meters (SLM), however, were expensive, big, and heavy.[12] To further stabilize this inscription device, the Acoustical Society of America and the American Standards Association issued the "American Tentative Standards for Noise Measurements" in 1936. This standard specified, among other things, reference pressure values,[13] frequency responses, and SLM microphone calibration procedure (Barstow 1940, Scott 1957). The IEC (International Electrotechnical Commission) classified different types of SLM based on accuracy.

The working premise of a current SLM remains pretty much the same. Sound pressure is captured by an omnidirectional condenser microphone and amplified. The signal passes through a frequency-weighing network. As equal-loudness contours vary depending on intensity (they become flatter as intensity increases), SLMs include the "A" weighting (based on the Fletcher-Munson contours at 40 phons)[14] and the "C" weighting, which roughly follows equal-loudness contours at 100 phons (the "B" weighting at 70 phons is now rarely used). As environmental noise levels tend to oscillate, the constant needle movement of early analog models made it difficult for the user to assign a value. For that reason, besides the frequency weightings, SLMs also include time weightings. The RMS (root means square) circuit provides the average energy of the measured sound in three standard durations: Fast ("F," with time constant of 125 milliseconds), Slow ("S," with time constant of one second), and Impulsive ("I," with 35-millisecond response). Finally, a logarithmic circuit converts the RMS circuit signal to deliver the reading in decibels. Digital SLMs include filters for real-time frequency analysis in bandwidths of one octave and one-third of an octave. A hold circuit stores the peak pressure (without the RMS averaging) and the maximum RMS values. Particularly useful for environmental noise measurements, the equivalent continuous sound level (L_{eq}) can replace time weighting by capturing noise levels sixteen times a second and calculating the average.

---

[11] Like the equal-loudness contour studies, the weighting procedure was developed by the telephone industry to measure residual noise in telephone circuits.

[12] H. H. Scott explains that some models could cost more than an automobile, weigh 110 pounds, and measure 4 cubic feet (Scott 1957, 1331).

[13] At 1 kHz, 0 dB = 20 μPa.

[14] 40 phons = contours at 40 dB at 1 Khz.

Package all this in a sturdy plastic case, and you have Ear 2.0, an actor in sound-politics that has quickly become the most authoritative representative of Ear 1.0 and, by extensions, of our ears. For the most part, the SLM is embedded in scientific discourse as a black box that can transport data without deforming it. In that sense, the SLM went from a source of *controversies* (how can we define the minimum thresholds, find workable units, and develop time and frequency weighting circuitries?) to a source of *resolutions*: a tool able to correlate the acoustic world "outside" with the bodily ear "inside." With SLMs, governments can hold a factory accountable for protecting its employees from noise—in most countries the limit varies between 85 dB(A) and 90 dB(A) for an 8-hour time-weighted average. Besides disciplining businesses, they can also estimate the damage caused by noise in a population. The U.S. Department of Labor states that each year 22 million workers are exposed to potentially damaging noise at work, with "$242 million [. . .] spent annually on workers' compensation for hearing loss disability."[15] In the 1990s, governmental agencies created the "disability-adjusted life-year" (DALY) to calculate disease burdens and life expectancy as the sum of years lived with disability and early deaths in the population. In a 2011 report on the quantification of healthy life years lost in Europe, the World Health Organization estimated that environmental noise had caused an accumulated loss of "61,000 years for ischemic heart disease, 45,000 years for cognitive impairment of children, 903,000 years for sleep disturbance, 22,000 years for tinnitus, and 654,000 years for annoyance [. . .]" (WHO 2011, v).

Drawing on Latour, one can envision in Ears 1.0 and 2.0 the constant encounter of two modes of existence. The first, the scientific one, operates through what Latour refers to as *chains of reference*, made of "points along the way that make it possible to verify the quality of our knowledge" (Latour 2013, 79). From laboratory to laboratory, developing more precise surgical instruments, dissecting animals, testing the ear through a series of stimuli, adding measurability, establishing a taxonomy, investigating disorders and diseases, documenting and averaging the organism's responses, all this has led them to a remote entity: the human ear. The constant extension and refinement of this chain allows the specialists to establish the series of correspondences between sound and ear. It is through chains of reference that both can be patiently constructed as knowable objects.

The performance of the SLM is informed by its capability to emulate Ear 1.0. To do that, it must translate a series of acoustic principles found in the human ear, such as the transduction potential of the diaphragm, signal amplification,

---

[15] U.S. Occupational Safety and Health Administration, https://www.osha.gov/SLTC/noisehearingconservation/.

weighting, and complex wave analysis. To provide a value that can be easily understood, stored, and compared, the SLM embeds the decibel. In its turn, the decibel has become part of a meta-language in acoustics to describe sound power (in lieu of Watt), sound intensity (in lieu of Watt per square meter), and sound pressure (in lieu of Pascal). However, as a unit that expresses the ratio between relative and reference values, the decibel can only operate under an agreement on what this reference value should be—in our case, the minimum hearing threshold. As this reference moved from telephone circuitry to environmental noise, from pure tones to complex waves, and from tests within one country to international surveys, it became easier to circulate and compare data and harder (but not impossible) to destabilize this value.

The second mode of existence to which the SLM belongs is that of technology, which entails the folding of actors to dislocate action. "With the folding of technological beings," Latour explains, "a *dislocation* of the action emerges into the world and makes it possible to differentiate between *two levels*, the starting level and the one toward which you have precisely shifted gears by installing in it other actors who possess different resistances, different durations, different degrees of solidity" (Latour 2013, 229). As the previous chapter has suggested and other sound studies authors have shown (Bijsterveld 2008, Mills 2011, Sterne 2003), the institution of the SLM, with its capacity to fold in its circuitry a scientifically reliable version of Ear 1.0, has produced a dislocation of action. We now delegate to the device the act of hearing for us. As it replaces our ears as the authoritative hearing actor, our ears become the *effect* of this technology. In other words, by depending on the SLM to assess any sound according to a limited set of criteria (frequency and amplitude), our hearing becomes conditioned by what the SLM—a black box whose resistances and degrees of solidity are different from the human ear—can hear. Through the miniscule repetition of a series of exposures to sounds that are allowed to exist thanks to the SLM's validation, this technological being is able to reshape our eardrums, cochleae, and hair cells. Following Latour, we can say that, by hearing for us, the SLM envelops not only the Ear 1.0's protocol but also the future of our acoustic world.

The ABNT's Hybrid Forum

Equipped with some knowledge about Ears 1.0 and 2.0, we can now enter a large meeting room in São Paulo, where roughly thirty people are staring at a projected screen. The tables in the room form a big "U," at the center of which the coordinator is editing the projected document as others jump in with comments and suggestions. On his side, the secretary types up the minutes. As the hours pass in what has been a long day of deliberations, the coordinator expresses some frustration with the group's slow conversion rate of collective knowledge into technical guidelines: "Let's try to write these sentences so we can express what we're thinking, okay? It seems that what's slowing us down is Portuguese grammar."

But what else is there to do? Once one has a reliable Ear 2.0, isn't the task at hand to simply point the device at a sound source (say, a vehicle or a jackhammer) and write the results down in a report? If only! Using an SLM is not like using a radar gun, which requires one to simply point at an object to estimate its speed by calculating frequency changes in the emitted radar signal or laser pulse. Capturing a sound can be quite challenging because, as a mechanical wave traveling from point A to point B, sound can undergo a range of diffractions and refractions depending on atmospheric conditions such as wind and temperature. Once the specialists have Ear 2.0 as a reliable translator of Ear 1.0, the device is ready to leave the acoustics lab and enter into messier acoustic conditions. That, in turn, requires them to design a *protocol* for properly measuring sound.

This is what the experts in the meeting room are doing. They are here voluntarily working to revise two Brazilian Regulatory Standards (*Normas Brasileiras Regulamentadoras*, or NBRs). The Brazilian National Standards Organization (*Associação Brasileira de Normas Técnicas*, or ABNT), which we came across briefly in the previous chapter, is the country's accredited standardization organization. NBR 10151 (1987, last revised in 2000) deals with the "Assessment of Noise in Inhabited Areas, Seeking the Comfort of the Community" by establishing "the required conditions for assessing the acceptability of noise in communities, regardless of the existence of noise complaints." The other standard, NBR 10152 (1987), addresses "Noise Levels for Acoustic Comfort" for different types of rooms, such as surgery rooms in hospitals, hotel rooms, living rooms, and concert halls. For several years, the acoustics experts have been trying to publish a revision of those two standards. But while most in the room agree that the current version of the standards is outdated, few seem to agree on what exactly the new documents should look like. Some want to make substantive additions, whereas others want to change as little as possible. In any case, something *must* be done because the task is well overdue—ABNT bylaws state that unrevised standards are canceled after ten years.

But why not simply use an already existent noise measurement standard? Why not take advantage of the work already done by the American National Standards Institute, the British Standards Association, the *Deutsches Institut für Normung*, the *Association Française de Normalisation*, or the *Instituto Argentino de*

Normalización y Certificación? What about the ISO, the International Organization for Standardization? Wouldn't the most practical solution be to just translate the international standard? Many Brazilian specialists argue that a national standard is relevant because it takes into consideration local circumstances. Local standards are important stimulants for a country's scientific research and economic development. Technical standards are not pure conveyers of technical conventions. They are necessarily related to economic concerns as well. For instance, as I mentioned in the previous chapter, construction practices in Brazil have differed from those of European countries and the United States for historical, climatic, economic, and safety reasons. Inserting construction standards from another country in Brazil could undermine the country's civil construction economy and favor foreign companies, which could increase even more the price of property in Brazil. Too much protection from external competition, however, can counteract any incentive for construction companies to standardize their buildings and improve quality.

The ISO claims that its mission is to "make products compatible [. . .], identify safety issues," and share "technological know-how and best management practices" (ISO 2016, 4). The organization was founded in 1947, in a period when the Global North was consolidating other initiatives of international cooperation, including the International Monetary Fund (1944), the World Bank (1945), the United Nations (1945), and the General Agreement on Tariffs and Trade (1947). In the 1950s, ISO published "recommendations for international use," most of which focused on screw threads, rolling bearings, freight containers, pipe sizes, couplings, and power transmission (ISO 1997, 35). In the 1970s, the organization started to publish international standards rather than recommendations. ISO has now expanded from mechanical standardization to a range of issues, including organizational management (ISO 9000), environmental management (ISO 14000), and social responsibility (ISO 26000).[16] For electric devices such as SLMs, the International Electrotechnical Commission (IEC, founded in 1906) is the main reference for technical standards.

According to a report published by the ABNT, "the creation of a national standardization organization [in Brazil] ended up connected to civil construction" (ABNT 2011, 45). The ABNT was founded in 1940 as a private nonprofit organization with the task of assimilating foreign procedures and developing guidelines for producing and testing materials (e.g., cement, concrete, paints, and elevators). In 1950, the federal government issued a decree determining the use of NBRs in

---

[16] The book that you now hold in your hands is equipped with an International Standard Book Number (ISBN), established by the ISO in 1970.

its projects (ABNT 2011, 56). By the 1970s, the organization's membership included a list of powerful private and public institutions, such as the Federation of Industries of the State of São Paulo, the Brazilian Development Bank, the São Paulo State Energy Company, the Brazilian Petroleum Corporation (Petrobras), the Civil Construction Industry Union (Sinduscon), and Eucatex (ABNT 2011, 58). The period of military authoritarianism in the 1970s brought a series of maneuvers to nationalize the ABNT. Although such attempts were unsuccessful, they led to the proliferation of other regulatory agencies.

Today, the ABNT is a small nonprofit private fish in a sea of governmental agencies. The broad network to which it belongs is the National System of Metrology, Standardization, and Industrial Quality (*Sistema Nacional de Metrologia, Normalização e Qualidade Industrial*, or Sinmetro), which integrates several state and private institutions related to standardization and quality control. These include the National Council of Metrology, Standardization, and Industrial Quality (*Conselho Nacional de Metrologia, Normalização e Qualidade Industrial*, or Conmetro), and the National Institute of Metrology Standardization and Industrial Quality (*Instituto Nacional de Metrologia, Normalização e Qualidade Industrial*, or Inmetro).

Conmetro, a normative agency made up of several technical committees, is responsible for defining public policies related to standardization. One of its committees, the Brazilian Standardization Committee, delegates to the ABNT the creation of national standards. Inmetro is Conmetro's executive arm. It participates in ABNT's committees and provides technical support to other technical committees, testing products and procedures (in accordance with the NBRs), and providing accreditation for other testing labs. This threefold institutional model (an overarching system, a normative and an executive arm) is present in other governmental sectors as well. For instance, in 1981, the Brazilian government created the National System of the Environment (SISNAMA), which includes the National Council of the Environment (CONAMA, the normative arm), and the Brazilian Institute of the Environment and Renewable Natural Resources (IBAMA, the executive arm).[17] Later in this chapter, we will see how some of these institutions are related to the revisions of NBRs 10151 and 10152. Without taking these governmental heavyweights into account, it is impossible to see why groups are so invested in the two standards.

Let us go back to the meeting room, which is taking place at the Sinduscon-SP[18] in downtown São Paulo. Here we already come across a first controversy. Why

---

[17] Another example would be the SNT (National Traffic System), created in 1997, which includes the National Traffic Council (Contran) and National Traffic Department (Denatran).
[18] Sinduscon was one of the earliest ABNT members.

are most of the ABNT meetings in São Paulo? Some participants defend having the meetings in São Paulo because it is a central location for those living in other regions and has a well-connected (and noisy, as we know) airport. Others complain that the choice of São Paulo is a strategy to keep the revisions closer to the construction sector and away from governmental agencies such as Inmetro and IBAMA, both of which are based in Rio de Janeiro. In 2012, when an expert from the state of Minas Gerais took over the coordination of the revisions of the two standards, the issue was raised once more, and it was decided that the meetings should be more inclusive and take place in other cities. As I show below, this change would soon cause turmoil.

The participants at the meeting are part of the of Acoustics Performance Studies Commission (henceforth the Acoustics Commission), one of the many commissions under the ABNT's Civil Construction Committee. As we take a seat at the table, someone hands over the attendance sheet. Beside columns for name and institution, the sheet asks participants to indicate their "class." This is not uncommon in standardization organizations, which operate thanks to volunteer work and aim to represent society more broadly by including experts, private organizations, state representatives, and laypersons. The sheet shows three classes of participants: "Neutrals," "Producers," and "Consumers." The categories seem to make more sense for the standardization of products such as lamps or screws, where it easier to distinguish manufacturing companies, research and governmental institutions that study and regulate lamps and screws, and groups that buy lamps and screws. For the Acoustics Commission participants (myself included), this classification was not always clear. Who exactly are the consumers? Due to this lack of certainty, and based on what I observed during the meetings, I propose that we organize the participants into five groups.

GROUP 1: SCHOLARS AND TECHNICAL INSTITUTIONS

This group is closer to the "Neutral" class. It includes university professors in departments of physics, civil engineering, electrical engineering, and architecture. It also includes representatives of testing materials/procedures institutions such as the IPT (see Chapter 1) and Inmetro. This group operates within research institutions and tends to push for accuracy, whether technical (the best measurement procedure), technological (the best equipment), or scientific (the best analytic protocol and terminology). The Brazilian Acoustical Society (SOBRAC, founded in 1984), which has served as an umbrella for acoustics scholars in the country, is an important institution attached to this group.

GROUP 2: LAW ENFORCEMENT AGENCIES

Standards express what experts in a given field agree to be the best practices. It is a recommendation, not an imposition. The weight of a standard, therefore, depends on how much legal power the government gives it. In 1990, CONAMA issued Resolution No. 1, which states that any noise above the limits established by NBR 10151 is to be considered harmful, and that construction projects should follow NBR 10152. We saw that, as a normative council, CONAMA deliberates practices that other agencies should follow. Through CONAMA's Resolution No. 1, the two NBRs have been embedded within numerous state and municipal noise laws. As the transcriptions of the meetings below make clear, the revisions would look different if CONAMA had not created this powerful link.

Group 2 includes members of the Environmental Agency of the State of São Paulo (CETESB), the municipal anti-noise agencies, IBAMA, and the State Public Prosecutor's Office. They are closer to the "Consumer" class because they deal with noise complaints through inspections, fines, and litigation. Whereas Group 1 is concerned with scientific precision, Group 2 is particularly attentive to the legal ramifications of the revised norms and the impact they will have on their modus operandi. To what extent, they ask, is it worth creating legal instability in the name of technical accuracy?

GROUP 3: SERVICE PROVIDERS

These are acousticians (mostly engineers and architects) in the private sector. They are invested in the revisions because it directly affects their jobs, which include conducting measurements, writing technical reports, and designing and implementing soundproofing projects. If Group 1 is concerned with scientific accuracy, and Group 2 with the standards' legal and budgetary ramifications, Group 3 sees these revisions as an opportunity to expand their business and consolidate an "acoustic mentality" in the country.[19] Together with Group 1, they believe the revisions can stimulate growth in the field of acoustics. One major challenge, however, is the lack of professionalization channels—in the early 2010s, the country had only one undergraduate program in acoustical engineering. ProAcústica (Brazilian Association for Acoustic Quality), which I include in this group, has been coping with this deficit by organizing several intensive courses in environmental acoustics. ProAcústica was created in 2010 and brings together acoustic

---

[19] We saw in the previous chapter that this "mentality" discourse has been floating around at least since the 1950s with the Brazilian Acoustics Institute.

engineers and soundproofing and construction companies—most of them based in São Paulo. As a result, ProAcústica should also be included in Group 4.

GROUP 4: PRODUCT PROVIDERS

This group includes companies that produce noise measurement hardware, software, and soundproofing materials. Like Group 3, they are directly affected by the revisions and expect them to increase sales. One of the main providers of SLMs in Brazil is 01dB,[20] the acoustics branch of the ACOEM Group—a French company founded in 1996 that focuses on environmental measurement, analysis, monitoring, and control. Nicolas Isnard, ACOEM's business manager in Brazil, explains that the development of SLMs took off in France in the late 1980s, with the stimulus of new legislation and standardization. 01dB has sold SLMs to several governmental agencies in Brazil, including the CETESB and the São Paulo Anti-Noise Agency (*Programa de Silêncio Urbano*, or PSIU). In the early 2010s, it was pushing for the introduction of acoustic simulation software to develop noise maps, a topic to which I return in the next chapter.

Another important actor in this group is São Paulo-based AtenuaSom,[21] one of the pioneers in Brazil in the soundproofed windows market. The company's biggest innovation is a technique for mounting a noise-canceling window onto an existent window. Recently, it started to test its windows using a holographic system, in which it is possible to visualize the soundproofing potential of different sections of window frame systems. Both 01dB and AtenuaSom are founding members of ProAcústica.

GROUP 5: NOISEMAKERS

These are private and public entities that participate in the revisions because their activities involve a certain amount of noise. Actors in this group include the Brazilian Association of Highway Concessionaires, the National Association of Railroad Transportation, the São Paulo Subway Company, the Federation of Industries of the State of São Paulo, the São Paulo Metropolitan Train Company, Vale S.A. (the largest producer of iron ore and nickel in the world), and the National Steel Company (*Companhia Siderúrgica Nacional*, a major steel producer). These actors are particularly interested in keeping the meetings in São Paulo, where they can monitor the revisions more easily.

---

[20] http://www.01db.com/pt-br
[21] http://atenuasom.com.br

I also include here Sinduscon and the Housing Union (Secovi),[22] both representatives of the construction sector. As we saw in the previous chapter, construction companies not only generate noise when assembling buildings but rely on questionable construction conventions when it comes to soundproofing. More recently, some companies started to see the improvement in their buildings' acoustics as a viable marketing strategy. They were particularly invested in another ABNT standard, NBR 15575 (published in 2013). This standard draws on ISO and establishes three noise transmission performance levels (minimum, intermediary, and superior) for construction systems (elevators, pipes, exhaust hoods, fans, etc.) in residential buildings. For Peter Barry, a senior expert at the IPT, "NBR 11575 is mostly about engineering. It deals with material performance, as opposed to NBRs 10151 and 10152, which deal with people" (personal communication, January 2017).

Table 2.1 gives a summary of the groups.

TABLE 2.1.

Five groups regularly present at the Acoustics Commission meetings for the revision of NBR 10151 and NBR 10152.

Group 1	Scholars, SOBRAC, Inmetro, IPT
Group 2	Law enforcement agents (CETESB, IBAMA, Public Prosecutor's Office, municipal agencies, etc.)
Group 3	Service providers (ProAcústica and acoustics companies)
Group 4	Product providers (01dB, AtenuaSom, etc.)
Group 5	Noisemakers (construction, traffic, mining, etc.)

In 2013, in a major push for the autonomy of the field in Brazil, members of the Acoustics Commission created the Acoustical Studies Committee, separating their activities (and the field of acoustics) from the Civil Construction Committee. They then tried to move NBRs 10151 and 10152 to the newly created Acoustics Committee. As the minutes of the meetings show, the Construction Committee strongly opposed this move, accusing the acousticians of a "lack of political awareness."[23] At what quickly became a tumultuous meeting, the Construction Committee representative explained that real estate companies and other "important entities in the country"[24] were deeply concerned about the impact that

---

[22] Although there is some overlap between the two organizations, Sinduscon includes construction companies, whereas Secovi is centered on real estate development companies.
[23] Minutes of the Meeting No. 11, NBR 10152 (Nov. 21, 2013), p. 4.
[24] Ibid.

the revised standards would have on urban planning. Using their administrative prerogative as the original holders of NBRs 10151 and 10152, the Construction Committee rejected the move and kept both standards within its territory.

In what follows, I consider six other controversies that emerged regularly during the meetings. These controversies illustrate the dynamics between the five groups and how specific issues moved back and forth between technical and social issues. I show this interaction by quoting exchanges between participants at the meetings I attended in 2012 (shortly after the Acoustics Commission meetings restarted with the new coordinator). This information is complemented with minutes from meetings I did not attend between 2011 and 2017. I do not identify the name of the individuals—just the group and institution he or she represents.

CONTROVERSY 1: PRESENT VS. ABSENT RECEIVER

> COORDINATOR (GROUP 3): *NBR 10151 is going to establish measurement procedures near the property. If the objective is to create a technical report about the noise impact on a residence, you need to measure noise at the façade of the building—that's what 10152 is about. So we will have these two possibilities, but it's not our responsibility to determine how the authorities are going to use these standards.*
>
> STATE PROSECUTOR (GROUP 2): *When you leave this undefined in the standard, the attorney is going to define . . .*
>
> COORDINATOR: *Yes, interpretations are going to arise. This is a technical standard. We need to predict both situations.*
>
> CETESB AGENT (GROUP 2): *The state prosecutor reads what the standard says and asks for things because the standard allows it. We are tired of being asked to measure things that are impossible to measure! You need to define in the standard that it does not apply to a given type of problem. The prosecutor makes us measure sound near a property, even when the community is 100 km away. So, you are going to penalize the property when there is nothing around it? You can't penalize a factory that is in the middle of nowhere.*
>
> COORDINATOR: *If you define an area for environmental preservation, you need to establish noise limits as well.*
>
> CETESB AGENT: *But you can make things stagnate that way. What about sugar and ethanol production here in the state of São Paulo?*[25] *If you apply this version of 10151 to measure noise limits, you are going to close all refineries!*

---

[25] Brazil is one of largest producers of ethanol fuel in the world.

METRO-SP REPRESENTATIVE (GROUP 5): *The point is not to give the opportunity to anyone to misuse the standard* . . .

ACOUSTICIAN (GROUP 3): *The 10151 needs to be a standard for outdoor environments to provide guidelines for projects, reducing noise impact and promoting sustainability. A factory can be in the middle of nowhere now. But urban planners might predict the emergence of a community next to industrial noise. We need to have this very clear. If we focus on the receiver only, we won't provide parameters for construction projects and urban planning. We will be just saying, "Let's wait to see what happens."*

METRO-SP REPRESENTATIVE: *The CONAMA resolution focuses on acoustic comfort, so it's about the receiver. The public sector should define the parameters and whether there will be urban expansion close to my traffic line project.*

ACOUSTICIAN: *Yes, the public sector will define the parameters based on the standard* . . .

COORDINATOR: *We already have twenty-three pages and there is a lot more to be done. This standard is going from four pages to almost forty pages. We are trying to let the debate flow here so we can have consensus.*

ACOUSTICIAN: *I suggest that we decide now if this standard is going to depend on the receiver or if it will address the noise impact regardless of the receiver. We need to settle this issue.*

COORDINATOR: *We will have infinite applications of this standard. It is important to make it cohesive, but some issues don't have to be decided now. We have very different situations in Brazil. We need to recognize CETESB's work against noise pollution here in São Paulo. Most of the topics discussed in this standard are obvious for us because we are experts. But for a layperson, the document is going to be the technical reference. We need to be careful.*

CETESB AGENT: *If you are going to build a residential building next to a noisy road, you must provide the necessary acoustic protection. Of course, that doesn't mean that the sound source can disrespect the limits. But whoever is arriving there afterward needs to anticipate protection measures because they are making use of that infrastructure.*

Part of the challenge here is that the Acoustics Commission was working on two complementary NBRs at the same time. Not only that, but the commission was expanding the documents considerably to include much more detailed measurement procedures and analytic criteria than the pre-existing versions. They were aware that such a move could risk allowing nonexperts, such as attorneys and lawmakers, to "misuse" the standard. The acoustician was pushing for a more proactive approach to the standards. Rather than focusing on the receiver as a parameter for defining sound pressure limits, the acoustician wanted the standard to become a determinant for the construction of hydroelectric plants, factories, highways, and other large development projects in the future.

This controversy relates to two different approaches to noise control: the passive approach, which deals with noise control based on the proximity to an existent receiver; and the active approach, which focuses on planning and establishes mechanisms for controlling noise regardless of the presence of a receiver. The active approach can have high implementation costs and is an effective tool for holding governments accountable for preserving acoustically sustainable spaces. An example of an active approach is that of noise maps, which describe sound propagation across an area and help define how much traffic or industrial activity that area should allow in the future. A good example of a passive approach is the São Paulo anti-noise agency, which, as we will see in Chapter 4, operates based on complaints.

CONTROVERSIES 2 AND 3: SCOPE AND TRAFFIC

> COORDINATOR: *The objective of NBR 15152 is to provide a measurement methodology. You measure at least three points in rooms with up to 300 cubic meters and additional points in larger rooms. In small rooms, one of the points measured needs to be in the corner. This is the methodology for diffuse acoustic fields.*[26] *You can measure with the windows open, windows closed, A/C on or off, and with the room empty or furnished. The professional is going to decide that, depending on the circumstances.*
>
> SCHOLAR 1 (GROUP 1): *So it is not "independent of sound sources" as the document says.*
>
> IPT REPRESENTATIVE (GROUP 1): *It is because we're measuring a room independently of any specific source. In the end, we are measuring the acoustic comfort—in those conditions and with those sources.*
>
> SCHOLAR 1: *This matter of environmental impact is a problem. NBR 10151 refers to external sources. NBR 10152 deals with noise coming from the building.*
>
> IPT REPRESENTATIVE: *You are measuring the overall noise in the environment. The source can be external, such as traffic, but you are not assessing the impact of traffic noise with 10152. You are simply assessing the comfort* inside *the room.*

---

[26] A diffuse sound field is a reverberant sound field in which the sound is more evenly distributed. A direct sound field, on the other hand, is sound in open space, without reverberation.

COORDINATOR: *If a train from the São Paulo Metropolitan Train Company is passing by, the purpose is not to assess its noise. It is important to highlight that this standard does not apply for environmental impact assessment.*

SCHOLAR 1: *But the environmental impact can be internal, too. We need to define that.*

SCHOLAR 2: *We could put "assessment of sound levels generated by sources outside the building."*

IPT REPRESENTATIVE: *The source doesn't matter!*

PROACÚSTICA ACOUSTICIAN (GROUP 3): *I think you are protecting the traffic infrastructure that way. This is a double-edged sword. You are going to have occasions in which the infrastructure will never be able to conform to NBR 10151. But you still could improve the soundproofing in the façade of the building that is exposed to traffic noise. It's much simpler than changing the traffic network.*

COORDINATOR: *We need a standard to assess and characterize a project, such as rail traffic. And we need a standard to assess and characterize an environment like the meeting room we are in right now, according to its use. These are two different things. We go back to that debate on responsibilities. Should the train company be held accountable? Or should the construction company that built the building be held accountable, knowing that the building is next to the railroad and it would be a meeting room? This needs to be clear. 10152 won't assess the impact of infrastructure. If during [sound] measurement, the train is a problem, I can't make the train company accountable. That's the construction company's fault.*

This exchange illustrates two heated controversies. The first one is the question of scope. As the same group of experts was revising two environmental noise standards at the same time, they were often unsure what each standard was supposed to do. The idea was to make NBR 10151 a standard for measuring the environmental impact of specific sound sources so that it could serve as a guide to urban planning. NBR 10152, however, was to be used simply to establish whether the noise level of a given room adhered to the limits for that type of space. The confusion was mostly about sounds coming from within the building. Which standard should be used to measure the noise made by a neighbor in the apartment above? After much discussion, the Acoustics Commission suggested that this issue could be resolved by framing NBR 10151 as a standard for measuring noise outside a given room instead of noise outside the building.

The second issue is traffic noise. In the previous chapter, we saw that traffic noise, including the minhocão and the Congonhas airport, has been one of the most contentious issues in São Paulo. We also saw that several noise ordinances provided a loophole for traffic. João Baring, one of the founders of the IPT acoustics lab and the coordinator of the Acoustics Commission in 2010–2011, considered traffic noise the most controversial point of the revisions. The strongest argument for excluding traffic noise from NBR 10151 was that, unlike stationary sources, which spread spherically, traffic noise spreads cylindrically. Measuring it using the procedures of NBR 10151 would yield incorrect results. At the same time, many experts argued that it would be absurd to leave out such a relevant source of noise pollution. This has been a delicate issue because NBR 10151 (2000) establishes that "if the background noise is higher than the numbers included in [the table] for the area and time considered, the criterion becomes the background noise." In other words, as the predominant and continuous source of noise in urban areas, traffic noise often became the reference value for measuring *other* noises.

In 2012, railroad, road, subway, and aircraft private and public organizations committed to creating standards for each type of traffic noise. Some experts found it questionable, however, that Group 5 would be in charge of drafting the standards for measuring their own noise. Others objected that, unlike NBRs 10151 and 10152, these new traffic noise standards would not have the powerful link to the CONAMA's Resolution No. 1/1990. In 2013, after heavy criticism, the Acoustics Commission decided to incorporate traffic noise into the standard by dividing NBR 10151 into two parts: one for general use (i.e., for inspections), and another for specific uses (air, waterway, rail, metro, and road traffic noise measurement). However, a few years later the Acoustics Commission changed its mind again, reverting to the idea of having separate standards for traffic noise.

CONTROVERSY 4: DECIBEL VALUES, PERIODS, AND ROOMS

> COORDINATOR (GROUP 3): *In large urban centers, we are recommending that daytime last between 7:00 a.m. and 10:00 p.m., and nighttime between 10:00 p.m. and 7:00 p.m. In small cities, we recommend between 7:00 a.m. and 8:00 p.m. for daytime, and 8:00 p.m. and 7:00 a.m. for nighttime.* [He reads the draft] *"For the application of this standard, nighttime must not last less than eight hours, must not begin after 11:00 p.m., and must not end before 6:00 a.m." That is to give some flexibility to industrial areas. Personally, I think the standard shouldn't include any timetable or decibel values. That is the responsibility of CONAMA, municipal governments, etc.*
>
> SLM COMPANY REPRESENTATIVE (GROUP 4): *Why does this standard include a timetable?*

IBAMA AGENT (GROUP 2): *We had a demand from the population. We were receiving complaints from Salvador in relation to carnaval sound trucks. The policing agencies didn't know what to do or what periods to establish. That was twenty years ago. . . . We needed to provide some parameters.*

INMETRO AGENT 1 (GROUP 1): *Don't change that table. When we have a law to create noise maps, then we can change the standard. Let's not forget that this is the only standard in Brazil that has the status of law!*

COORDINATOR: *I really don't think we should put this to a vote . . .*

INMETRO AGENT 2: *We did not go to most meetings in São Paulo. You would put this to a vote if we were absent, and now you can't put it to a vote because the others are absent?*

COORDINATOR: *Previous meetings had more than thirty people on average. The idea was: since we are changing the standard, let's fix what's broken. This is broken. We are assessing two long periods, daytime and nighttime, recognizing the possibility that municipalities could use an additional time band . . .*

IBAMA AGENT 1: *Did anybody oppose the 10:00 p.m. daytime limit in previous meetings?*

COORDINATOR: *I did. The industrial sector needs some flexibility . . .*

INMETRO AGENT 1: *That is a political decision! If you change the time, you are going to penalize the lower classes, people who have to wake up early to take buses and drive across the city.*

[Tumult; several people talking at the same time]

COORDINATOR: *The logic is: you make the change. If society disagrees, they are going to say so, and we go back to how it was.*

INMETRO AGENT 2: *No! The opposite makes more sense. If society wants the change, we make the change. You should only change it if there is some technical justification.*

COORDINATOR: *There is no technical justification for nighttime to be from 10:00 p.m. to 7:00 a.m. The commission made that change. We will write this standard from scratch; it's not the same standard.*

INMETRO AGENT 1: *You can't make a standard for a few Brazilians only!*

COORDINATOR: *We are only saying that if a municipality establishes the time periods, nighttime shouldn't be less than eight hours and shouldn't start after 11:00 p.m. and end before 6:00 a.m. This document was written in São Paulo.*

INMETRO AGENT 2: *A controversial issue like this should have been put to a vote.*

COORDINATOR: *It's only controversial for you! If it were controversial to anyone else, it would have been voted on, and it would be in the minutes. All I'm asking is . . . Let's leave it like it is and see what happens in the national consultation.*

I'm really worried about making this change. That is why I asked to put in the minutes that the consensus was to keep this part as it was decided in São Paulo. That way I protect myself, too . . .

The NBR 10151 noise limit/daytime/zone table has been perhaps *the* most contentious issue among the Acoustics Commission participants. Many experts agree with the coordinator that a standard should not include noise limits and timetables and should leave that issue to municipal noise ordinances. Then again, as the Inmetro agents are quick to point out, the CONAMA resolution gives these standards considerable legal weight; leaving the table out at this point could generate legal instability. The IBAMA official, who was involved in the creation of the first version of NBR 10151 in 1987, explains they included the table in the standard because they were receiving several noise complaints and inspection agencies did not know what parameters to use. At previous meetings in São Paulo, the Acoustics Commission had reached an agreement about making some changes to the timetable to give cities some flexibility. However, to the coordinator's dismay, the consensus suddenly evaporated when the meeting took place in Rio de Janeiro, where IBAMA and Inmetro are headquartered. One particularly contentious issue about the NBR 10151 table was the noise limits in rural areas (see Table 2.2), which

TABLE 2.2.

NBR 10151/2000 decibel values suggestion for daytime and nighttime according to area type. All values in dB(A).

Area Types	Daytime (7:00 a.m.–10:00 p.m.)	Nighttime (10:00 p.m.–7:00 a.m.)
Farms and ranches	40	35
Strictly residential urban areas; areas near hospitals and schools	50	45
Mixed-use, predominantly residential	55	50
Mixed-use area, with commercial and administrative potential	60	55
Mixed-used area, with recreational potential	65	55
Strictly industrial areas	70	60

Group 5 considered too low and an obstacle to industrialization in Northern and Center-West Brazil.

NBR 10152 also includes a table. It establishes a range of decibel values acceptable for different room types (e.g., hospital nurseries, hotel restaurants, and school libraries). But while some in the Acoustics Commission wanted to make the NBR 10151 table more flexible, others wanted to make the NBR 10152 table more restrictive. "The construction company lobby is very powerful," explained a senior acoustician who had followed the unsuccessful attempts of previous revision commissions to change the NBR 10152 values. A particularly contentious issue related to this table is that, in 1990, the Ministry of Labor issued the Ministerial Order 3751, which refers to the NBR 10152 table to determine noise exposure limits in the workplace.

CONTROVERSY 5: MEASUREMENT

COORDINATOR (GROUP 3), READING THE NBR 10151 DRAFT: *"The distance between the microphone and any reflective surface besides the floor needs to be at least twice the distance between the microphone and the sound source's dominant surface." Should we keep this? That is from ISO 1996 . . .*

IPT REPRESENTATIVE (GROUP 1): *That's because when the sound passes through the microphone the propagation is more uniform, more spherical.*

COORDINATOR: *Correct. . . .*

IPT REPRESENTATIVE: *But that depends on the frequency. I suggest we establish two conditions: either measuring two meters away from any surface or with the microphone on the façade. At two meters, there is no correction, and on the façade, you take out 6 dB. Intermediate positions increase the level of uncertainty, which can be a problem for inspection.*

SCHOLAR 1 (GROUP 1): *To make things easier, we could include the possibility of measuring one meter from the window, so that the technician would be able to stretch his arm out the window.*

SCHOLAR 2 (GROUP 1): *But who has a one-meter arm?*

SCHOLAR 1: *Everybody!*

COORDINATOR: *We excluded the measurement procedure from NBR 10151 in which you could measure inside the complainant's room. That procedure was criticized because it was subject to the room's acoustic field.*

CETESB AGENT (GROUP 2): *We from CETESB oppose changing the measurement procedure. Which side are you on? We defend the people; the individual who cannot enjoy his property because there is a noisy factory. The ISO standard* was designed for European buildings that have soundproofing. By changing the procedure, you are removing the policing power from the state agencies. That is extremely important for us because we will not be able to satisfy any complainant by measuring noise outside his room.

COORDINATOR: *This is a commission with representatives from all parties concerned. We need to find a balance to reconcile these issues in order to avoid privileging one specific sector. We all will have to adapt.*

SCHOLAR 1: *I understand CETESB's concern in the sense that this change can make a difference subjectively. Complainants like to see the inspector taking measurements where the nuisance takes place (in their bedroom, in the office, etc. . . .).*

Measurement procedures were another highly contentious topic. NBR 10151/2000 includes the procedure where the technician measures sound pressure levels in the complainant's room. According to the standard, the technician should subtract 10 dB(A) from the values in the table when measuring inside the complainant's room with the windows opened. Experts from Groups 1 and 3 argued that measuring an external sound inside the room was too unreliable due to reverberations. The idea, then, was to limit 10151 to noise measurements outdoors, where sound propagation is more diffuse. Group 2 strongly opposed this change, claiming that measuring outside the building would create a problem when the noise source was inside the building (e.g., a bar or nightclub on the ground level).

After much pressure from Group 2, the Acoustics Commission decided to leave in the option for measuring inside buildings with the windows open and subtracting 5 dB(A) from the table values in case of structure-borne noise. This, in turn, led to another disagreement on how the technician would be able to distinguish between airborne and structure-borne noise during an inspection. Another disagreement was the type of SLM the document should require. Groups 1, 3, and (obviously) 4 wanted to require more accurate devices (with frequency-spectrum analysis) and more regular calibration than suggested in the existent standards. Group 2 reacted strongly, claiming that this would be too costly for inspection agencies and require additional training.

CONTROVERSY 6: VOCABULARY CHANGES

SENIOR ACOUSTICIAN (GROUP 3): *I'm worried about nontechnicians using this standard. This is a revision of a previous standard. The previous standard included "acoustic comfort" in its title; technicians and nontechnicians understand what it is about. When we change the title to "sound pressure level,"* nontechnicians won't understand that it's about acoustic comfort. We should find a way to link the old standard with the new.

SCHOLAR (GROUP 1): *In the 1970s and 1980s, I was in favor of making super-simple standards, almost like an instructional booklet, for easy assimilation. But now the nature of it has changed.*

SENIOR ACOUSTICIAN: *People are going to look for a noise standard and are not going to find it. It's "sound pressure level" now, not "noise."*

SCHOLAR: *Those interested in assessing noise in any environment, indoors and outdoors, will have to study the topic.*

SENIOR IPT TECHNICIAN (GROUP 1): *In discussions about performance standards, the word "comfort" was the most controversial one. This is more common in the thermic field, where you have measurements based on satisfied people. They thought discussing comfort was complicated because it is a technical standard. Acousticians adopted that approach, too.... We can add "comfort" somewhere in the body of the standard, but it shouldn't be the focus. The focus is that there is a parameter, and we are going to measure it and have criteria to face up to it. That is the concept of "technique."*

STATE PROSECUTOR (GROUP 2): *Attorneys and lawyers use the word "comfort" because that is the major demand today in Rio de Janeiro. We have to understand that this will cause an impact. There are several ongoing lawsuits related to comfort.*

SCHOLAR: *I'm against "acoustic comfort." It's too subjective.*

One final controversy worth mentioning is the proposition to eliminate the terms "noise" (*ruído*) and "acoustic comfort" (*conforto acústico*) from the standards. The experts wanted to remove "subjective" and "nontechnical" terms. Because of this general concern, NBR 10151's title changed several times during the revisions. Its 2000 version, "Evaluation of Noise in Inhabited Areas Aiming for the Comfort of the Community," was later replaced with "Measurement and Evaluation of Sound Pressure Levels in Outdoor Environments." With the discussions about indoor measurements (Controversy 5), the title changed to "Measurements and Evaluation of Sound Pressure Levels in Inhabited Areas." NBR 10152 went from "Noise Levels for Acoustic Comfort" to "Measurement and Evaluation of Sound Pressure Levels in Indoor Environments."

Although the senior acoustician believed that *ruído* should be maintained in the title because it relates to the English word "noise" and is easier to understand, others insisted that it was too subjective. Many seemed eager to make the standards the mark of a new era of professional acoustics in Brazil, where those dealing with noise measurement "would have to study." They argued that the only way to improve the acoustics of Brazilian cities was to consolidate a workforce capable of generating accurate and stable facts. From now on, technicians should refer to the SLM as a *sonômetro* rather than the popular *decibelímetro* because the device doesn't measure decibels, but rather sound pressure levels. By 2016, it seemed clear that the field of standardization in acoustics had grown considerably in the country. The newly created Acoustics Studies Committee was overseeing several workgroups to revise, resurrect, translate (from the ISO), and create a wide range of standards about sound barriers, terminology in acoustics, indoors acoustics, musical acoustics, acoustic treatment, audiometry, and electroacoustic instrumentation. This suggests that the experts have in fact moved from noise measurement procedures into broader debates concerning Ears 1.0 and 2.0.

Like other standardization organizations, the ABNT operates via consensus. Once the experts reach an agreement about a standard, the draft then goes to public consultation online, where the broader public is invited to provide feedback and vote either for or against its confirmation. After analyzing the votes, the ABNT decides whether to send the document back to the Commissions for further revisions. Since 2013, the Acoustics Commission has submitted different versions of the two standards for national consultation, but each time it has failed to receive enough favorable votes. By 2017, only NBR 10152 had successfully made through the national consultation.

Michel Callon et al. (2009) define as "hybrid forums" public spaces where a heterogeneous group of actors meets to decide the best options for the collective across different fields of expertise. This process entails the conception of a state of the world inhabited by human and nonhuman entities. The authors argue that science is incapable of defining all possible worlds, simply because many uncertainties prevent a group of experts from identifying all entities at play—and hence all outcomes. The controversies at the ABNT meetings discussed here show the challenges of "establishing a clear and widely accepted border between what is considered to be unquestionably technical and what is recognized as unquestionably social [as] the line describing this border constantly fluctuates throughout the controversy" (Callon et al. 2009, 25).

Which actors in the Acoustics Commission get to define where this fluctuating border lies? Which entities should be included in the normalization of noise measurement procedures? What state of the world do they want to create through temporary agreements officially inscribed in audiometric charts, microphone circuitries, and technical standards? Should the standard *passively* describe or *actively* prescribe a given acoustic environment? Should the experts recognize the standards' ramifications within political circles, legal channels, and administrative flows, or should they stick to rectified knowledge of scientific chains of reference? The ABNT debates highlight the connections between hybrid forums and sound-politics. Meeting after meeting, the Acoustics Commission moved back and forth between the technically accurate and the politically and legally relevant. In trying to mobilize the revisions in a certain way, each group strategically highlighted specific attachments between the document and the world "out there."

As the frameworks provided by science and technology are not definitive, the question becomes how to deal with the overflows created by the hybrid forums. "All, specialists included, think they have clearly defined the parameters of the proposed solutions, reckon they have established sound knowledge and know-how, and are convinced they have clearly identified the groups concerned and their expectations. And then disconcerting events occur" (Callon et al. 2009, 28). The decision of whether to include traffic and construction noise is an example of overflow, in which the barriers containing a technical framework fall apart. Both construction and traffic noise groups have used their economic and political weight to break away from chains of reference linking their activities with noise control measures. At the same time, note that not even Ears 1.0 and 2.0, the building blocks of noise measurement procedures, are definitively stabilized. They too go through their own overflows each time new studies suggest existing technical inaccuracies and further revisions in equal-loudness contours and SLMs.

At the Acoustics Commission, the traffic and construction sectors entered the meetings and secured their places in the decision-making process by creating a collective and organized voice. The construction sector, which has been central in the history of standardization in the country, has maintained institutional control over the two standards. At the core of the country's construction hub, and away from Brasília and Rio de Janeiro (two national political centers), the governmental agencies struggled to participate in the meetings and were usually a minority. When that issue overflowed, and it became impossible not to include those noises in the technical standards, they made sure the specialists understood the political implications of their techno-scientific endeavor. The traffic group, on the other hand, reversed the strategy and lobbied for an even *more reliable* chain of reference, claiming that the propagation of sound in traffic is different from stationary sources and would thus require specific measurement procedures.

NBR 10151 will continue to lurk throughout the book. Lawmakers will refer to it to change or propose laws, law enforcers will apply it when inspecting bars, and judges will quote it to justify their sentences. Entangled in other controversies centered around other modes of existence and accessing these two technical standards from afar, other groups assume the NBR 10151 and 10152 are solid black boxes that can transport information between our human ears and the SLM screen seamlessly. However, as this chapter showed, under closer examination, the terrain of standardization in which the experts navigate is fairly slippery and highly mediated. If, as a modernist project combining different lines of inquiry (from physics, medicine, and telecommunication), New Acoustics has provoked a crucial shift in noise control debates, establishing and securing the chains of reference and technological folds that allow governments to do sound-politics remains open to heated discussion.

> The silent lobby of the well-dressed men can have more impact than all the noise made by shirtless protesters.
>
> GUILHERME BOULOS, leader of the Roofless Workers' Movement, "A Batalha do Plano Diretor" (Boulos 2014).

3

THE ECHO CHAMBER

APRIL 29, 2014, 2:00 P.M. It is a sunny afternoon in downtown São Paulo as I walk into a giant ear (see Figure 3.1). Passing through the ear canal, I see the eardrum and the small bones responsible for transmitting and amplifying sound waves. As I arrive at the inner ear to observe the cochlea in its magnified detail, two people walk toward me. They ask if I would be interested in learning more about the ear. Walking me through it, they explain that the human ear is a delicate instrument, capable of detecting the widest range of stimuli of our sensory apparatus. Noisy environments make us susceptible not only to hearing loss (the destruction of irreplaceable hair cells), but also to cardiovascular disease and sleep disturbance. To remedy this invisible threat, the two specialists argue, we need efficient laws and a broad public campaign to raise awareness among city residents. As we walk out of the ear canal, I notice a sound level meter on a tripod connected to a laptop (see Figure 3.2). The experts explain that the device is measuring environmental noise and sending data in real time to a website so that Paulistanos can monitor the dangerous noise levels they are exposed to daily.

I leave the giant ear and walk toward the São Paulo Municipal Chamber (henceforth the Chamber), where the "First Municipal Conference on Noise, Vibration, and Sound Disruption" is taking place. Created thanks to the initiative of councilor Andrea Matarazzo and with the sponsorship of several public and private institutions, this three-day conference intends to raise awareness about noise

FIGURES 3.1 AND 3.2. Inflatable giant ear in front of the São Paulo Municipal Chamber in April 2014. Next to the giant ear, the SLM mounted on a tripod monitors the environmental noise (the results are displayed in real time on the monitor).
Credit: *Câmara Municipal de São Paulo*.

pollution in São Paulo with informed discussions about noise-control technology, legislation, and quality of life issues.

While the specialists prefer to focus on scientific, technological, and legal aspects of noise, many members of the audience take the opportunity to complain about the inefficacy of São Paulo's Anti-Noise Agency (*Programa de Silêncio Urbano*, or PSIU) and the lack of conviviality among some Paulistanos. One elderly man vents his frustration by listing the nuisances he is exposed to every day, especially the "nerve-racking" explosive sounds of motorbikes with modified exhausts. Others are interested in the relationship between noise and crime. Are noisier cities more violent? Does noise stimulate crime? Although the panelists of acousticians and audiologists deny any direct causal relation between noise and violence, all seem to agree on the influence of environmental noise on stress, which in turn can trigger violent behavior. Many in the room recall a 2013 incident in an upmarket gated community, in which a businessman shot and killed his neighbors over disagreements about TV volume.[1]

As the event progresses, the sounds of drumming and shouting start to penetrate the conference room. Members of the Roofless Workers' Movement (*Movimento dos Trabalhadores sem Teto*, or MTST), camped out in front of the Chamber, are pushing the councilors to vote for approval of a new Master Plan. For weeks, the press, politicians, nonprofit organizations, and scholars have been discussing this piece of legislation. Mayor Fernando Haddad, from the center-left Workers' Party, wants the plan to tame real-estate speculation and increase public housing close to central areas. The noise created by protesters outside the Chamber is a political instrument; although hours earlier they were removed from the Chamber's benches for disturbing the proceedings, they can still make their presence known through noise. Resonating through the windows and walls inside the Chamber, their sound is opening up the politics of shared existence.

For these protesters, the new Master Plan is an opportunity to change the city's well-known spatial segregation, where residents of the few modern, wealthy and car-centric districts with good infrastructure experience the city very differently from the majority living in poor, nonregulated and mostly nonwhite peripheries. The protesters are aware that a group of councilors who disagree over some items on the bill is threatening to block the vote. Around 5:00 p.m., the Chamber leadership decides to postpone the vote because internal review commissions have failed to submit their assessment in time. Frustrated, some protesters burn tires and try

---

[1] "Briga entre vizinhos resulta em 3 mortes em condomínio de Alphaville." *G1* (May 23, 2013), http://g1.globo.com/sao-paulo/noticia/2013/05/briga-entre-vizinhos-resulta-em-3-mortes-em-condominio-de-alphaville.html.

to gain access to the Chamber. In a few minutes, the soundscape changes when the police's stun grenades replace the protesters' noise. The quiet that follows, accurately registered by the sound level meter next to the giant ear, has a lot to say about noise, space, and citizenship in Brazil.

In this chapter, we are going to visit the Chamber to examine three noise debates in São Paulo, the first two of which involve noise ordinances created in the 1990s[2] and enforced by the PSIU. Entering the public sphere at a moment of transnational anxieties about city violence and the effects of human activity on the environment,[3] both ordinances introduced changes in São Paulo's everyday social relations and generated heated discussions about urban noise. The first debate revolves around the Evangelical lawmakers' attempts to exclude, minimize, or hinder the impact of the noise ordinance on religious services. The second debate focuses on an ordinance that requires bars without acoustic insulation to close at 1:00 a.m.; a demand that faced strong opposition from nightlife businesses. The third debate circles back to the beginning of this article. I describe the recent attempt of a group of acoustic engineers to lobby the city administration for the systematic mapping of traffic noise. For the sound specialists, it is only with such an acoustic map that the municipal government can plan a truly sustainable city.

Noise Laws

Brazil's 1988 Constitution is celebrated as a landmark for its attempt to relate urban development and citizenship issues. As Leonardo Avritzer describes, new political parties such as the Workers' Party and the Brazilian Communist Party, and non-governmental organizations (NGOs) and other civil groups such as the National Movement for Urban Reform, were deeply active in the National Constituent Assembly, demanding more participative models of city organization (Avritzer 2007). The Constitution requires all cities with more than 20,000 inhabitants to have a City Master Plan. The Master Plan should include public hearings and stimulate the participation of city residents. In 2001, urban citizenship gained new impetus in Brazil when the Statute of the City was enacted. The Statute requires cities to create urban policies in order to "guarantee the right to sustainable cities, understood as the right to urban land, housing, environmental sanitation, urban infrastructure, transportation, public services, work, and leisure for current and

---

[2] Laws 11501/1994 and 12879/1999.
[3] The Rio de Janeiro Earth Summit (1992) and the Kyoto Protocol (1997) point to the growing transnational concern with environmental issues in the 1990s.

future generations."[4] Such policies include requiring owners to use their property according to its social function, allowing "residents of small urban housing lots to obtain original ownership title if they can prove five years of continuous residence without legitimate opposition" (Holston 2008, 292), raising property taxation for owners of underutilized land, and selling additional building rights to private landowners to generate revenue for public housing projects.

Besides the insurgent citizenship of marginalized and working-class Paulistanos, the impressive growth of Evangelical organizations was equally important for the configuration of a new citizenship paradigm in the country. Brazil's Evangelical population has grown from 6.6% in 1980 to 22.2% in 2010,[5] and is today one of the largest in the world. The Universal Church of the Kingdom of God (founded in 1977) and the Church of the Grace of God (1980) are the most influential groups, operating at all levels of political administration. The so-called Evangelical Caucus entered the political landscape during the National Constituent Assembly to ensure religious freedom and keep a check on the historical alliance between the state and the Catholic Church. Since then, the defense of Evangelical interests and values "is supported not with the Bible, but with the Federal Constitution" (Trevisan 2013, 588). Through their emotional fervor and religious devotion, Evangelical cults have imprinted an ethos of collective life away from "the pure and simple struggle for daily material survival" (Corten 1999, 57).

In the following sections, I come back to these issues. For now, two points need to be stressed. First, the "noise" of the MTST in front of the Chamber described above is one instantiation in a long process of urban citizenship in post-dictatorship Brazil, marked by demands for popular participation and political accountability. As Gabriel Feltran notes, "This nascent public sphere offered a sounding board for the new social movements, particularly those of popular inclination, which could then be seen as the 'new actors' in the Brazilian political scene" (Feltran 2007, 85). Second, although discussions about housing, public transportation, land rights, and religious practices may seem unrelated to urban noise, the simple fact that these matters are mediated by municipal policymakers converts these issues into forces within a local field of political bargaining. In Brazil (and Latin America more broadly) public urban problems such as noise tend to galvanize political action via insurgent citizenship (struggles for access) and differentiated citizenship (conservation of privileges) claims discussed in the Introduction.

---

[4] Article 2 of the Statute of the City, Law 10257/2001.
[5] "Número de evangélicos aumenta 61% em 10 anos, aponta IBGE." *G1* (June 29, 2012), http://g1.globo.com/brasil/noticia/2012/06/numero-de-evangelicos-aumenta-61-em-10-anos-aponta-ibge.html.

Three main pieces of legislation address environmental noise in Brazil at the federal level. Article 42 from the 1941 Penal Contravention Laws (PCLs) establishes as a misdemeanor "disturbing someone's peace or work by (I) shouting or causing uproar, (II) conducting noisy or annoying activity, in contravention of the legal prescription, (III) using sonic instruments or acoustic signals abusively, (IV) provoking or not preventing noise made by animals under one's guardianship." The article does not specify the duration, location, or intensity necessary to characterize these infractions, which makes the document difficult to use.

Similarly to when dealing with a misdemeanor in the United States, citizens rely on the police to enforce Article 42, which generates another problem. When a disturbed noise complainant calls the police to get rid of an unwanted sound, they need to go to the police department to fill out a police report. Residents are usually unwilling to do this for fear of retaliation. Once the police officers leave, they feel unprotected. Second, within the sphere of civil litigation, the Brazilian Civil Code (2002) includes a section on neighborhood law. It protects property owners against "interferences that are harmful to the safety, peace, and health of the inhabitants, caused by the use of neighboring property,"[6] in which case "the neighbor may require their reduction or elimination whenever possible."[7]

The third federal protection against noise is the more serious Law 9605/1998, known as the Environmental Crime Law. Submitted by the executive branch in 1991, one year before the United Nations Conference on the Environment and Development in Rio de Janeiro (the 1992 Earth Summit), the bill languished for seven years in the House of Representatives and Senate before arriving on President Fernando Henrique Cardoso's desk for approval. Article 54 of the law establishes as a crime "To cause pollution of any kind at levels that result or may result in damage to human health, or cause the death of animals or significant destruction of flora."[8] This is the closest connection to noise pollution in the document. Interestingly, the bill that was voted and approved by the Senate did include a clause prohibiting the production of "sounds, noises, or vibrations at odds with the legal or regulatory provisions, or disregarding the norms on emission of noise resulting from any activity," with a punishment of three months to one year in jail. In 1998, Cardoso vetoed this clause, arguing that the PCLs had already tackled the problem. The veto was the result of powerful lobbying by the Evangelical caucus in the Senate, and Cardoso later admitted that his decision was made "out of respect

---

[6] Brazilian Civil Code, Article 1277.
[7] Ibid., Article 1279.
[8] Law 9605/1998, Artigo 54.

for religious freedom."⁹ We will return to these three laws in Chapter 5, when considering possible channels of noise-related litigation.

Whereas many Brazilian cities rely on the PCLs and the Environmental Crime Law to tackle noise complaints, major urban centers like São Paulo have developed specific laws and institutions to fight noise, often attaching them to zoning laws. In Chapter 1 we saw that São Paulo has passed several laws regulating noise. The 1955 ordinance stimulated the conception of the city into zones and helped to separate industrial noise from the citizens. The 1974 ordinance drew on the city's first comprehensive zoning laws and was created a few years after city residents became exposed to the noise pollution of the minhocão expressway and the Congonhas airport jetliners. However, despite the potential of these ordinances, the continuous lack of law-enforcement seemed to suggest that noise regulation in São Paulo was simply unattainable.

This perception changed in the early 1990s, when the city administration, together with councilor Roberto Tripoli (one of the founders of the Brazilian Green Party in the 1980s), created a new noise ordinance for the city. According to Regina Macedo, Tripoli's longtime assistant, the councilor suggested to the newly elected Mayor Luiza Erundina, from the Worker's Party, the creation of a task force to outline a new noise ordinance.¹⁰ The mayor accepted the suggestion and assembled a group of specialists in 1991, nominating Tripoli as the coordinator. After analyzing the data at hand, the specialists suggested four major strategies for tackling the issue. First, they argued that it would be necessary to create regular campaigns to raise awareness about the "physical and psychological harm that noise pollution causes to human beings."¹¹ Second, it would be necessary to create an agency to centralize noise complaints and conduct regular inspections, measuring noise levels and fining offenders. Third, the executive branch should create a mechanism to require sensitive public buildings (such as schools and hospitals) to be properly soundproofed. Finally, the municipal government should focus on localized problems, especially noise complaints related to neighborhood noise, including bars, restaurants, churches, fitness centers, and other commercial establishments.

In 1992, Tripoli marched to the City Hall with forty residents directly affected by neighborhood noise to give visibility to the issue and pressure his colleagues in the Chamber to recognize its importance. Wearing earplugs, the protesters handed over three hundred noise complaints to the Municipal Secretary. On International

---

⁹ "Lei Ambiental é sancionada com 10 vetos," *Folha de São Paulo* (February 13, 1998), 6.
¹⁰ Personal communication, October 2012.
¹¹ Regina Macedo, "Poluição sonora: histórico de uma luta de seis anos do vereador Roberto Tripoli," unpublished dossier (n.d.).

Noise Awareness Day in April, Tripoli installed a soundproofed booth in downtown São Paulo. For two weeks, curious Paulistanos were invited to enter the booth and hear silence in the middle of the downtown hubbub and become aware of the amount of noise they were exposed to regularly. At that point, the councilor had already collected more than four hundred noise complaints from residents.

According to Helena Sobral, Erundina's environmental advisor, the task force looked at other noise ordinances, especially the New York City Noise Code. "I remember that the New York City noise ordinance required dog owners to submit their dog to devocalization," recounts Sobral, "and drivers were allowed to honk only in emergency situations. We didn't want to go that far."[12] According to Sobral, the idea of creating a project to tackle noise pollution in the city came not from the councilor, but from the mayor, as a maneuver to give visibility to her political allies in the Chamber. As Sobral explains,

> The mayor was having some trouble getting support from the environmentalist sector in the Chamber, and that is why Tripoli was invited as a coordinator. In a certain way, behind the scenes, this bill was a way for the mayor to approach the Green Party. There was a movement in the Workers' Party that argued that there should be stronger control of environmental issues—that there should be some kind of agency taking care of that. The issue was sensitive because Rio 92 [the Earth Summit] was getting close.[13]

Political opponents portrayed Luiza Erundina as an inexperienced socialist unfit for managing Latin America's largest metropolis. As a member of the Workers' Party whose election represented a significant victory, and facing the enormous challenge of putting into practice the demands of lower-class citizens, Erundina encountered a hostile Chamber. Contentious issues between her administration and council members included the regularization of squatters, progressive property taxation, approval of the yearly budget and public transportation fares. In 1991, with only 21 out of 53 councilors on her side, the mayor began to seek alliances with other parties.

In the rationale for the bill, Tripoli stated that: "Noise pollution, which according to the World Health Organization is already the third-most-serious type of pollution affecting humanity (behind air and water pollution), has grown at an alarming rate in São Paulo."[14] The document argued that localized noise from bars

---

[12] Personal communication, July 2013.
[13] Ibid.
[14] Bill 707/1993.

and churches kept residents from sleeping and resting properly. It also provided a list of physiological and psychological problems related to noise pollution, and mentioned its economic downsides, since "the individual affected by noise pollution has less capacity to concentrate and less motivation to work."[15] Clause 3 of the bill 707/1993 proposed that:

> Venues intended for leisure, cultural activities, lodging, recreation or religious services, which can be adjusted to standards similar to residential standards or which can allow the setting of special noise and vibration standards, will have to provide acoustic treatment that limits the passage of sound to the exterior, if these activities use sound sources with live transmission via amplifiers.[16]

Such commercial venues would need to get a specific license based on the type of activity, sound equipment used, the zone where the property was located, opening hours, and the maximum noise levels allowed. Noisy venues would need to provide a technical report signed by at least two acoustic engineers, showing that the place had been properly soundproofed. The license would have to be renewed every two years. Venues without a license or with an expired license would: a) receive a fine of 300 UFMs, roughly US$12,000[17] for the first offense; b) face closure for the second offense, followed by administrative closure with the sealing of all entrances and confiscation of sound equipment for the third offense. Venues with a valid license caught making noise above the legal limits would: a) receive a fine of US$2,000 (venues with capacity for up to 50 people), US$4,000 (up to 100 people), US$6,000 (up to 200 people) and US$8,000 (more than 200 people); and b) be subject to the same procedure as venues with no license for second and third offenses. If the commercial establishment continued to fail to abide by the law, the police would help inspectors to close it again. If they continued not to abide by the law, the fines would increase to US$12,000 per month.

In 1993, Tripoli submitted the bill to the Chamber for voting. At this point, Erundina and the Workers' Party had passed on the administration to Paulo Maluf.[18] Still, the passage of the bill was smooth sailing. The bill got 76% approval

---

[15] Ibid.
[16] Ibid.
[17] The UFM ("Municipal Fiscal Unit") is a unit used for local taxes and fines; its value fluctuates according to the average of daily or monthly interest rates negotiated by banks. Taking the conversion rate of March 2016 (US$1 = R$3.60), 1 UFM = approx. R$144, or US$40. Values of fines shown in this book use this currency conversion rate.
[18] Maluf over-invoiced several construction projects and diverted public funds to his accounts overseas. In 2007, the Manhattan district attorney's office indicted Maluf, claiming that US$140 million had passed through his secret account at Safra National Bank in New York between 1997 and 1999. In 2012, a court

in the first voting session and 83% in the second. In March 1994, the Chamber sent the document to Maluf, and in April it was signed into law. Clauses in Tripoli's law delegated responsibility to the municipal government for creating an anti-noise agency resourced with the "mechanisms to manage complaints and [. . .] inspect and measure noise levels."[19] In that same year, São Paulo's Anti-Noise Agency, the PSIU, was created. In 1995, Tripoli submitted a revision of his law to allow venues to remain open (respecting the noise limits) after the first two offenses while they adjusted to the law. In 1996, the original Law 11501/1994 was updated to Law 11986/1996.

RELIGIOUS NOISE

In the Chamber, the most contentious issue in Tripoli's noise ordinance was the inclusion of religious services. Merely two months after it passed into law, Evangelical councilor Gilberto Nascimento submitted a change to the document, arguing that the reference to "religious services" went against Article 5 of the Constitution, which establishes that "Freedom of conscience and creed are inviolable, and the free exercise of religious cults is granted with the guaranteed protection, in the form of law, of the places of worship and their rites." Nascimento also argued that the acoustic measures required by the ordinance would have a detrimental effect on the churches' architectonic design, part of São Paulo's "cultural heritage." Still, although the amendment was successfully approved,[20] it did not technically prevent the PSIU from fining churches.

Councilors Celso Cardoso and Dito Salim, in 1997 and 1998 respectively, were less successful in their attempts. The former proposed explicitly removing churches from the scope of the ordinance in order to "protect [. . .] a highly-needed right of those that congregate to externalize the most diverse liturgy-related manifestations in our country."[21] The latter submitted a bill that added obstacles to enforcing the ordinance against churches: from 10:00 p.m. on, the sound of religious services could be measured only after ten residents (from different households) had complained. The complainants' full names, addresses, and identities would have to be provided. As Salim explained:

---

on the Island of Jersey found him guilty of stealing US$10.5 million by issuing over-inflated invoices for the construction of an eight-lane highway during his administration.

[19] Law 11501/1994, São Paulo.
[20] Law 11631/1994.
[21] Bill 1010/1997.

According to the will of the majority of the population and all Evangelical and Catholic communities, it is necessary to give flexibility to the municipal legislation that addresses urban noise, in order to allow religious temples of any nature to practice their worship or religious services with singing, praising, clapping, and other characteristic noises, without applying the sanctions of this law. [. . .] Specialists in urban noise unanimously claim that constant uninterrupted noise is the type of noise harmful to health. This is not what happens in religious temples. [. . .] Establishing a minimum of ten complainants prevents one or two people opposed to a given religious creed activating the municipal policing machine and its harsh fines for personal reasons, against the will of hundreds or even thousands of people. [. . .] In the world we live in today, I am positive that the more prayers, worshipping, and praising there are, the less pain and anguish there will be among the suffering population of this immense metropolis.[22]

After failing to get both bills approved by the Chamber's internal review commissions, Evangelical lawmakers decided to try another strategy. In 2001, Marta Suplicy, from the Workers' Party, took office as mayor after beating Paulo Maluf in the second round. Suplicy, who during the elections had claimed that 35 of the 55 councilors were nothing but thieves,[23] would later encounter difficulties in the Chamber as mayor. She wanted to pass a bundle of eleven bills that would allow her to, among other things, create new offices, open new *cargos de confiança* (literally "positions of trust") and increase property taxation.

To get the bundle approved, Suplicy's allies in the Chamber started to reach out for support. The Workers' Party leader in the Chamber openly stated that those who supported Suplicy's bundle would be able to "suggest bills of interest to them."[24] The group of ten Evangelical councilors saw this as a unique opportunity to push for a more enduring legal shield for their churches against noise control. "I have to help the Workers' Party so they can help me,"[25] Evangelical councilor Carlos Apolinário stated bluntly. Indeed, his vote was decisive for passing Suplicy's bills in a divided Chamber (see Figure 3.3). Shortly afterward, Apolinário submitted a bill establishing that the PSIU would be able to measure the sound of

---

[22] Bill 740/1998.
[23] "Câmara tem 35 bandidos," *Folha de São Paulo* (August 27, 2000), A20.
[24] "Acordo garante apoio a projetos de Marta," *Folha de São Paulo* (June 26, 2001), C7.
[25] Ibid.

FIGURE 3.3. A 2001 political cartoon with the title "Give, and it will be given to you," depicts an exchange between Evangelical councilor Carlos Apolinário and mayor Marta Suplicy. Apolinário: "In exchange for our support, you ease the PSIU for us, okay? Hallelujah, sister!" To which Suplicy responds: "You might not know, but deep down you are a little PT [Workers' Party]."
Credit: *Folha de São Paulo* (September 18, 2001), C5.

churches only from inside the complainant's property, with proper identification from the complainant plus three external witnesses, who would accompany the measurement in the complainant's house (remember that residents are usually unwilling to expose themselves to potential retaliation). In case of irregularities, the churches would have ninety days to comply with the law and soundproof the building. Additionally, fines could not exceed US$140 (R$500), even in the case of repeated infractions.

A few months later Apolinário changed his bill, replacing the fixed US$140 fine with a proportional penalty: fine of US$250 for places with occupancy of up to 500 people, up to US$2,222 for churches with occupancy of 5,000 people or more. In practical terms, since it completely excluded the threat of administrative closure, the bill would allow churches to make noise above the limits if they wanted as long as they paid the fine—much more affordable than under Tripoli's law. Apolinário argued that his bill was based on the constitutional guarantee of the "free exercise of worship" and was a response to the necessity of "regularizing the inspections which are often conducted in a scathing and discretionary manner."[26]

The Chamber's commissions quickly reviewed the bill and issued a collective one-page document supporting it. The bill caused much dissent inside and outside the Chamber, starting with Suplicy's own party. Many Workers' Party councilors accused the bill of benefiting the churches and openly stated that they would vote against it. During discussion of the bill, Apolinário and Tripoli had to be held back by colleagues to avoid physical aggression.[27] Pressured by partisans, the mayor indicated that she would veto the bill—a move that, according to Apolinário, would turn Suplicy into an "enemy of the Evangelicals."[28] In the end, the mayor approved the document, and Apolinário's controversial bill passed into law.

A few years later, São Paulo State's attorney general filed an injunction to suspend Apolinário's law, on the basis that the Constitution considered *all* noises equal before the law. Apolinário then proposed another bill, simply transferring the content from his now-defunct law. However, to avoid problems of constitutionality, he replaced the word "church" with "meeting places for 100 people or more." The bill maintained that measurements would be carried out from inside the complainant's property, accompanied by witnesses; fines would range from US$140 to US$2,222 (depending on the building capacity); the PSIU would have to wait thirty days to register a new complaint; and venues would have ninety days to comply. Once again, Apolinário drew on the Constitution, explaining that "any exorbitant economic sanction would inhibit the social activity inserted in and inherent to meeting places, thus conflicting with the Federal Constitution."[29]

Again, the municipal chamber quickly approved the bill and sent it to the mayor's office. In 2008, however, Mayor Gilberto Kassab vetoed the bill, arguing that it contradicted the zoning laws, which established limits of noise levels in the city to be measured in front of the venue and not in the complainant's property. Besides, Kassab stated, Tripoli's law already tackled the problem of noise pollution in the city. The mayor also drew on the Constitution, claiming that all places are equal before the law: bills should not make a distinction between meeting places and other types of places. In 2010, the Chamber voted to reject Kassab's veto and passed the bill without his approval. The press suggested that such disregard for the mayor's authority was a strategy of some Chamber members to pressure the administration to offer more positions to politicians affiliated with the Evangelical

---

[26] Bill 203/2001.
[27] "'Favor' pode reduzir multa por ruído," *Folha de São Paulo* (September 18, 2001), C5.
[28] "Marta vetará projeto que beneficia templos," *Folha de São Paulo* (September 27, 2001), C6.
[29] Bill 399/2007.

churches. In 2010, the State Attorney General once again intervened and suspended Apolinário's law.

This analysis of noise control debates in the Chamber suggests how Evangelical lawmakers have attempted to exempt church services from noise control measures. Tensions between religious practices and noise control are of course not particular to Brazil or to Evangelical churches. In his analysis of legal decisions concerning sounds of worship in the United States in the nineteenth and twentieth centuries, Isaac Weiner shows the intricate ways in which Protestant and Episcopal church bells, Jehovah Witnesses' preaching and Muslims' calls to prayers have been contested in court. Weiner argues that framing such sounds as a nuisance and "acoustic seizure" draws on post-Enlightenment notions of "immature faith, overly concerned with external behavior rather than interiorized commitment and insufficiently respectful of the rights of others" (Weiner 2014, 6). Besides showing how religious groups embed moral values in the public sphere, these case studies unveil how lawyers and judges articulate legal documents to shift the soundscape of cities.

Like other Pentecostal congregations around the world, Brazilian Evangelical churches deploy sonic "excess" as part of a cathartic experience. Contrasting with Catholic services (still dominant in Brazil), salvation is here conceived of more as a process than as a state. As Corten notes, full and complete membership in Pentecostalism involves "declaring in a loud voice, in the heart of the neighborhood community, that one 'accepts Jesus [as Savior]'" (Corten 1999, 27). Fervor is sustained, guilt externalized, and evil expelled throughout the service with praising, hand-clapping, foot-tapping, and stomping. Glossolalia, the verbal utterance of incomprehensible words pervasive in Pentecostalism, "represents a mysticism in which one has knowledge of, a revelation of, and an immediate contact with God, in noise instead of in silence" (Corten 1999, 100).

Evangelical religious practices in Brazil rely on what Martijn Oosterbaan refers to as "sonic supremacy" across the urban fabric (Oosterbaan 2009). This supremacy involves as many as five daily services (amplified to the limit of the speakers' power) and extensive use of radio and TV. As sensory submersion sustains the edifice of religious devotion, it has been crucial for the politicians affiliated with the Evangelical churches to protect them against noise control. In following the tortuous journey of Tripoli's bill, we can observe how the balance between sound and politics shifted as the document passed through different hands, attaching itself to discourses and legal documents as actors tried to reframe what place (if any) religious "noise" should have in a noise ordinance. Evangelical politicians have used a wide range of strategies to allow their churches to resonate in the city, changing or making new laws, closing political deals and ignoring the Mayor's right to veto.

However, despite their prominence in the legislative debates, religious sounds are not the primary source of noise complaints in the city. For the majority of Paulistanos, the problem has been bars. Of the 496 noise complaints Tripoli collected between 1992 and 1997, 252 (50.8%) were related to nightlife noise.

NIGHTLIFE NOISE

Physician Jooji Hato was a member of the Chamber from 1982 to 2011. In 1996, he proposed a bill to prohibit bars from staying open after 1:00 a.m., arguing that the measure would decrease the chance of a "series of problems caused by the citizen who stays the entire night in these places drinking alcohol."[30] According to Hato,

> Bars here operate differently from those in Europe. When London pubs toll those bells at 11:00 p.m., that's it—no more alcohol.[31] Here youths drink all night long, often ending up in a coma from alcoholic intoxication, and bar owners see that and don't do anything. That's what I was seeing when working in the emergency room.[32]

One Chamber review commission approved Hato's bill due to the wide support it had received from the population. Since cities in Europe and the United States closed bars early to reduce alcohol-related crime and accidents, cosmopolitan São Paulo should follow their example. The Finance and Budget Commission, however, did not support the bill because it would have detrimental effects on the city's economy. The commission explained that closing bars at 1:00 a.m. was unconstitutional since the state should not "prohibit or prevent licit economic activities."[33] The Bar and Restaurant Union claimed that, if it passed, the ordinance would lead to the loss of up to 120,000 jobs in the city.[34] Bar owners explained that Brazilians go out late, so half of the profits made during a night occurred between midnight and 3:00 a.m. The Chamber commission also argued that "Happiness needs to be externalized. São Paulo is a city where people work a lot, much more than in any other city in the country. Where would the leisure of our citizens take place? How can we remove our citizens' right to chat until the time they want?"[35] The

---

[30] Bill 396/1996.
[31] In fact, many venues in British cities have special permits that allow them to operate after 11:00 p.m. For a discussion of debates on laws related to nightlife activity in the UK, see Hadfield (2007).
[32] Personal communication, December 2012.
[33] Finance and Budget Commission, Assessment of Bill 396/1996.
[34] "Toque de recolher começa na madrugada," *Folha de São Paulo* (July 14, 1999), 4.
[35] Finance and Budget Commission, Assessment of Bill 396/1996.

commission thus resorted to a well-known modernist discourse about São Paulo to highlight the importance of leisure (noise) to counterbalance the city's strong work ethos. A major point of contention was Vila Madalena, historically a residential neighborhood that in the last decades had become a bohemian nightlife center among middle-class youth.

Realizing that the original document was too draconian in the country's current cultural and economic climate, and yielding to external pressures, Hato decided in 1998 to revise the bill. Venues would be able to stay open after 1:00 a.m. if they followed three rules: had proper acoustic insulation, operated with completely closed doors, and provided a parking lot and security personnel. Drawing on Tripoli's Law, he established the punishment of US$12,000 for the first offense, closure for the second, and administrative closure with use of judicial force if the bar defied the second-offense closure. The Bar and Restaurant Union's director estimated that less than 1% of the 40,000 venues potentially affected by the bill were soundproofed.[36] The press suggested that the bill was in fact introduced in the Chamber to generate "political noise," taking attention away from the series of corruption charges brought against Mayor Celso Pitta.

Pitta took office having the majority in the Chamber, but his term was anything but smooth. The first black mayor to be democratically elected in São Paulo, Pitta won the elections as Maluf's protégé and was elected thanks to the support of his political godfather. Already in his first year as mayor, he faced a parliamentary inquiry commission concerning his involvement, as Maluf's finance secretary, in influence peddling and embezzlement of public funds. The scandal increased when the mayor severed relations and exchanged spiteful accusations with Maluf. In four years in office, Pitta had his personal bank account blocked and was removed from office twice. To avoid having the Chamber open yet another parliamentary inquiry commission, the mayor proposed to "open a new communication channel"[37] with a group labeled by the press as "rebellious" councilors. The group was part of Pitta's political coalition in the Chamber but threatened to vote against him if the mayor continued to ignore their demands—Pitta's estranged wife claimed that the mayor bribed the councilors to get the necessary support. Hato, himself mentioned in a police inquiry of a bribery scheme at community health centers, was one of the "rebels."[38]

In 1999, notwithstanding the controversies and thanks in part to Pitta's backers in the Chamber and a survey showing that 67% of Paulistanos approved the

---

[36] "Pitta vê 'com simpatia' projeto sobre bares," *Estado de São Paulo* (June 24, 1999), C4.
[37] "'Rebeldes' promovem outra CPI contra Pitta," *Estado de São Paulo* (April 16, 1998), C7.
[38] "Polícia investiga 8 vereadores que derrubaram CPI," *Estado de São Paulo* (October 1, 1999), C3.

document,[39] Hato's bar bill passed into law with 38 votes in favor and 13 against. After being signed into law as Law 12879/1999, the document continued to generate controversy, particularly because the ordinance did not provide a definition of "bar." Restaurants, nightclubs, and even bakeries have bar areas selling mostly alcohol. Should they be included in the law? When pushed for a definition of "bar," the São Paulo Secretary of Supply and Sanitation told reporters to ask an engineer, who would have "better technical skill" to give a precise definition of a bar.[40] A few weeks later, the Hotels, Restaurants, and Bars Union were granted an injunction establishing that the law would apply only to "bars, excluding those venues that do not commercialize solely alcoholic beverages."[41] Following that, and to make matters even more confusing, bars started to include food and nonalcoholic beverages on their menus to circumvent the law. In 2001, councilor João Antônio submitted a bill to get restaurants and cafes off the hook, arguing that doing so would prevent an increase in unemployment and a negative impact on São Paulo's vibrant culture. A few years later the bill was passed into law as Law 13772/2004.

As indicated above, and as he would later explain in a book defending his noise ordinance (Hato 2010), Hato's main concern was to limit alcohol consumption to reduce the number of accidents.[42] He also maintained that the law would lead to a reduction in serious crime such as homicide, particularly in the suburbs. According to the councilor, a study conducted by the University of São Paulo in 1996 had shown that 48% of crimes in São Paulo took place in and near bars, motivated by impulsive behavior.[43] By establishing a strong causal link between urban disorder and crime, Hato was following the zero-tolerance premise. A controversial public security policy, zero tolerance became popular in the 1990s during Rudolph Giuliani's administration in New York City, which Hato often referred to when promoting his bill—despite the fact that New York has a zone-based ordinance for bar closures rather than a city-wide one. Giving more autonomy to precinct commanders, Commissioner William Bratton justified such actions as part of the NYPD's "dual emphasis on quality-of-life signs of crime as well as on serious crime" (Bratton 1998, 36). According to the commissioner, zero-tolerance policies contributed to the city's economic development by creating safer urban spaces and helping the city to revitalize. Critiques of zero tolerance have pointed out that such approaches have overpopulated the prison system, over-militarized the

---

[39] "Sanção foi um evento político," *Folha de São Paulo* (July 14, 1999), 4.
[40] Quoted in "Bar é o que vende bebida, diz secretário," *Folha de São Paulo* (July 16, 1999), 3.
[41] Quoted in "Liminar derruba lei dos bares em SP," *Folha de São Paulo* (July 16, 1999), 3.
[42] According to Hato, Brazil's powerful beer industry lobbied aggressively against the bill (personal communication, December 2012).
[43] "Câmara aprova fechamento de bares à 1h," *Estado de São Paulo* (June 23, 1999), C1.

police, and antagonized minority groups. In the following chapters, we will revisit the zero-tolerance premises.

Between 1999 and 2006, the homicide rate per 100,000 inhabitants in São Paulo dropped from 69.1% to 23.7%, more than in any other city in the country. The most impressive difference observed in homicide rates took place in the city's poor suburbs. Some argued that, like "dry laws" in other places, Hato's law had a positive role in decreasing homicide. Others, including councilor João Antônio, insisted that the law had negative effects because it increased unemployment and, consequently, the city's jobless population. Still others, such as criminologist Paula Miraglia, contended that "there is no consensus about what could have motivated the reduction of homicide [. . .] in the city" (Miraglia 2011, 339).[44]

Hato's emphasis on alcohol consumption and violence reduction might suggest that environmental noise was only incidentally related to this debate. This is not the case, however. Already before it became law, São Paulo state's adjunct-secretary of public security stated that the police would not interfere with its enforcement because most supporters of the bill had mentioned nightlife noise and not public security issues.[45] There was also some contradiction between the law and available data. The same 1996 University of São Paulo study used by Hato to defend the bill showed that the majority of homicide incidents occurred between 10:00 p.m. and midnight rather than after 1:00 a.m.[46] Shortly before sanctioning it into law, Pitta conceded that the bill's scope had changed, as nightlife noise became the major justification for its existence.[47] In 2001, Mayor Suplicy issued a decree making the PSIU responsible for enforcing Hato's 1:00 a.m. Law.[48]

Whereas the first debate went from a noise issue to a question of religious freedom, the second debate started by focusing on violence and moved on to noise. The indiscriminate location of bars in the city with disregard for zoning laws, and the wide popular support against such venues, made the specificity of noise less of an issue than the religious sounds in Tripoli's law. In debating the latter, many Evangelicals refused to accept a noise ordinance that put together nightlife noise and religious sounds. For them, the former was part of a hedonistic cult of spiritual weakness, a generator of dysfunctional families populated by evil forces ready to turn into selfish addicts those who frequented bars. Hato,

---

[44] Some authors suggest that the decrease in homicides in the late 1990s is directly related to the rise of the Primeiro Comando da Capital (First Command of the Capital, PCC), a highly organized crime faction that has been able to regulate crime and violence in the city suburbs. See, for instance, Willis (2009).
[45] "Pitta deve sancionar fechamento de bar à 1h," *Estado de São Paulo* (July 7, 1999), C3.
[46] "Pitta quer 'cooperação' do Estado para fiscalizar bares," *Estado de São Paulo* (June 25, 1999), C4.
[47] "Pitta deve sancionar fechamento de bar à 1h," *O Estado de São Paulo* (July 7, 1999), C3.
[48] Decree 40798/2001.

who endorsed Apolinário's bill, argued that religious sounds are better than nightlife noise because they represented something positive in the community.[49] Yet, beneath the hubbub of religious services and bars, traffic noise continued to resound, heard by all but opposed by few.

TRAFFIC NOISE

As discussed in Chapter 1, São Paulo became an important economic hub in the late nineteenth century, largely because of the expansion of its coffee industry. This helped to boost investment in infrastructure, which in turn stimulated the establishment of manufacturing industries. As rents in central districts went up and new public health laws prohibited dense agglomerations, poor residents gradually left the central districts and migrated to the peripheries. Using the argument that it would be too costly to expand the existing rail lines in order to provide public transportation to the relocated working classes, the municipal government opted to allow private bus companies to operate in those areas. By the mid-1960s, tens of thousands of bumpy roads connected suburbs to industrial districts and the city center. As São Paulo embraced sprawling modernity, traffic noise (from trams to buses and aircraft) became a constant problem. However, this has commonly been framed either as an unavoidable aspect of urban life or as a private issue[50] rather than a responsibility of the public administration.

Traffic noise is what those who mounted the giant inflatable ear in front of the Chamber in 2014 wanted the city administration to address. "The municipal administration is rigorous with bars and restaurants [. . .], but it does not do anything to prevent the noise coming from the streets, particularly that generated by buses and trucks,"[51] explained acoustic engineer Davi Akkerman, then president of ProAcústica. The organization found a point of entry to the Chamber with councilor Andrea Matarazzo, then president of the Chamber's Urban Affairs Commission. In 2013, Matarazzo passed a resolution instituting annual conferences on noise control and pollution, to be organized and promoted by the Chamber. The purpose of the event was to bring together residents, academics, the private sector, and politicians to discuss and raise awareness about environmental noise, and to "create the guidelines for an effective legislative and administrative undertaking against urban noise."[52]

---

[49] Personal communication, December 2012.
[50] Since the 1990s business in soundproofed windows has boomed.
[51] '"Rigor com bares não vale nas ruas,' diz especialista em acústica," *Folha de São Paulo* (May 4, 2014), 12.
[52] Resolution 18/2013.

ProAcústica wanted lawmakers to include a section requiring the development of acoustic maps for the city in the new City Master Plan, under discussion in the Chamber. They argued that only with this tool would it be possible for the government to establish a reliable diagnosis of the problem and to plan the city properly, according to the noise it generates. Noise maps took off in the early 2000s in Europe, when European Commission Directive 49 (2002) determined that member states should harmonize permissible noise levels, paying special attention to "road and rail vehicles and infrastructure."[53] Member states would work to preserve quiet areas by using strategic noise mapping to orient the development of action plans. Additionally, authorities should make noise maps accessible to the public. Noise maps require state-of-the-art software to provide a reliable visual representation of traffic noise levels across space, indicating zones of excessive noise in the city and facilitating the development of strategies for mitigating noise pollution. In Brazil, the city of Fortaleza was the first to generate a comprehensive noise map to tackle noise pollution. One of the consultants of the Fortaleza noise map was Bento Coelho, a Portuguese acoustician who had been invited by ProAcústica to come to Brazil seasonally to develop courses to increase the workforce in acoustics.

Traffic noise is a crucial component of environmental noise in Brazil because it is the background against which other noises are measured. As mentioned in the previous Chapter, NBR 10151 establishes that if background noise is higher than the values included in the table, then the reference should be the background noise. Thus, areas with intense traffic are less likely to have trouble with noise control regulation. According to Nicolas Isnard, the representative in Brazil of acoustics equipment company 01dB and one of the founding members of ProAcústica,

> The moment Brazil is going through is similar to what happened in Europe ten years ago. It is a moment of mapping out and measuring urban environments so we can have an idea of what is actually going on. You only understand the problem when you know the sound sources and how the sounds propagate. If you focus on a specific sound source, you will remove that source but another one will appear.[54]

In 2013, Matarazzo submitted a bill proposing the creation of a noise map in São Paulo within twelve months after its passing into law. It would draw on the guidelines of zoning laws (sanctioned by Suplicy in 2004), include public hearings,

---

[53] European Union, "Directive 2002/49/EC—25 June 2002," *Official Journal of the European Communities* (2002), 12.
[54] Personal communication, June 2012.

and be carried out by the city's environmental quality department. Arguing that the noise of airports, highways, and railways has serious consequences for the city residents' health, the bill entrusts to the city administration responsibility for creating quiet zones, using new technology to reduce noise, and setting deadlines for the reduction of sources above noise limits. In 2014, Matarazzo also submitted an amendment to the Master Plan law requiring the creation of a noise map within twelve months. In common with the two debates discussed earlier, the noise map raised controversies. However, unlike Tripoli's and Hato's bills, resistance came not from civil society, but from the City Hall.

In 2014, a few months after the MTST street protest described at the beginning of the chapter took place in front of the Chamber, the City Master Plan was approved. Sympathetic to the demands from landless and working-class activists (who helped get him elected in 2012), Workers' Party Mayor Fernando Haddad doubled the area allocated for public housing and created mechanisms to expropriate idle land used for real-estate speculation. Haddad vetoed the noise map amendment, arguing that noise is "intrinsically mutable and dynamic, which in itself denotes the infeasibility of such endeavors by the executive branch."[55] Workers' Party councilors also argued that twelve months was not enough time to create the map, adding that the controversy could delay the creation of zoning laws, which are legally attached to the Master Plan. Matarazzo then changed the time frame in his bill, giving the administration four to seven years (depending on the area) to generate the noise map. In 2016, Haddad signed Matarazzo's Noise Map bill into Law 16499. In 2017, another project addressing traffic noise was submitted to the Chamber. Bill 405/2017 makes explicit the city administration's responsibility for minimizing air and noise pollution generated by public transportation and freight vehicles. Referring to the 2016 Noise Map Law as an important advancement to improving quality of life in the city, the bill includes annual mandatory noise inspections of buses and trucks according to the CONAMA resolutions (discussed in Chapter 2).

In 2016, Haddad approved the controversial zoning laws,[56] which will define for the next decade what can be built and what type of activities carried out in the city spaces. Assuming Paulistanos abide by the document, São Paulo will have more commerce corridors in residential areas, more public housing, higher buildings next to major thoroughfares, and more people using public transportation. Some residents believe the city will get noisier; not (only) because it still lacks a noise map, but because the zoning laws turned several residential areas into mixed-use

---

[55] Law 16050/2014.
[56] Law 16402/2016.

zones (which have higher noise limits) and gave churches more lax noise limits.[57] Articles 146 and 147 of the new zoning law replace the Tripoli's Noise Law and Hato's 1:00 a.m. Law, respectively. For Article 146, the law establishes roughly US$2,770 (R$10,000) for the first fine, US$5,540 for the second fine, and US$8,310 plus administrative closure for the third fine. For Article 147, it establishes roughly US$2,220 (R$ 8,000) for the first fine, US$4,440 for the second fine, and US$6,660 plus administrative closure for the third fine.[58]

The press argued that, in lowering the fine, the new law benefited noisy groups such as Evangelical churches. When a new mayor took office in 2017, this group saw an opportunity to obtain even more advantages. This is not surprising: as we saw throughout this chapter, the entrance of a new mayor sets in motion a series of negotiations with the Chamber to ensure the administration will be able to pass sensitive laws somewhat smoothly. In 2017, the religious caucus (which included Evangelical and Catholic leaders) included an amendment to a law submitted by the mayor to regulate the payment of municipal debts, including fines. The amendment establishes amnesty for accumulated debts in fines of up to US$33,330 for each religious temple. As Evangelical councilor Eduardo Tuma explained, "Our intention is to do justice to an entity that assists the state, that does not aim at profits and ameliorates society, as in the case of assisting drug users."[59] In July 2017, mayor João Dória (PSDB) signed into law[60] what many labeled a shameful "amnesty" to the churches' illicit acoustic behavior.[61]

NOISE CIRCLES

Latour describes politics as the process of "ceaselessly retracing one's steps in a movement of *envelopment* that always has to be begun again in order to sketch the moving from a group endowed with its own will and capable of simultaneous freedom and obedience" (Latour 2013, 135). Unlike the scientific ideal of retrieving "facts" through straight chains of reference, and technology with its folds, political action moves in circles. It does so because it turns around issues as they become

---

[57] On weekends and during holidays, churches can operate between 6:00 p.m. and 10:00 p.m. with the highest noise limits (7:00 a.m.—7:00 p.m.) authorized for each zone.
[58] This book focuses on the period before the replacement of the 1:00 a.m. Law and the Noise Laws.
[59] Quoted in "Por pressão da 'Bancada Cristã,' SP aprova anistia de dívidas e multas às Igrejas," *G1* (June 22, 2017), http://revistapegn.globo.com/Noticias/noticia/2017/06/pegn-por-pressao-da-bancada-crista-sp-aprova-anistia-de-dividas-e-multas-as-igrejas.html.
[60] Law 16680/2017.
[61] "Câmara de SP aprova anistia de dívidas de IPTU e multas às igrejas," *O Estado de São Paulo* (June 22, 2017), http://sao-paulo.estadao.com.br/noticias/geral,camara-de-sp-aprova-anistia-de-dividas-de-iptu-e-multas-as-igrejas,70001857118.

matters of concern, oscillating between unity and multitude, representation and obedience:

> Start with a multitude that does not know what it wants but that is suffering and complaining; obtain, by a series of radical transformations, a united representation of that multitude; then, by a dizzying translation/betrayal, invent a version of its pain and grievances from whole cloth; make it a united version that will be repeated by certain voices, which in turn—the return trip is at least as astonishing as the trip out—will bring it back to the multitude in the form of requirements imposed, orders given, laws passed; requirements, orders, and laws that are now exchanged, translated, transposed, transformed, opposed by the multitude in such diverse ways that they produce a new commotion: complaints defining new grievances, reviving and spelling out new indignation, new consent, new opinions. (Latour 2013, 341)

To move politically is to continuously start over this impossible circle, at the risk of letting it disappear and ending up with restless or indifferent multitude rather than a unified "we." It is not hard to see how noise engages with the constant renewal of the political circle. If, as suggested in the Introduction, noise raises ontological and spatial debates as it propagates, isn't that movement *in itself* the delineation (however ephemeral) of a political circle already? To hear a noise is to be surrounded (encircled) by it. But what *is* it? Does it belong *here*? If it does not, then eliminating that noise becomes a matter of pushing the noisy entity *outside* my acoustic field. But maybe I am not alone in this task. Maybe there is a multitude willing to eliminate this noise. The multitude becomes unity by turning around the issue. If this new "we" fails to neutralize the noisy entity by persuading it to move away from our newly created circle, perhaps we can invite the lawmakers (who represent an even larger circle) to turn around our issue with us. But then, of course, the noisy entity might have its own representatives too! In that sense, lawmaking in sound-politics always includes at least three circles: the acoustic field propagating in concentric waves, groups converting these waves into a political issue by turning around it (thus creating unity out of the multitude), and representatives inserting themselves in these circles either to solve the issue or to mobilize it for other (more self-serving) reasons.

As residents call upon the city administration to regulate noise, the matter becomes entangled with political struggles and administrative skirmishes. Owing to its political heterogeneity and culture of bargaining with the stronger executive branch, the São Paulo Chamber is a particularly advantageous point from which to examine the relations between city noise and citizenship. The three debates examined here suggest that the Chamber is a powerful mediator of sound production in the city. Like other public controversies, noise enters the Chamber not as a self-contained problem with reachable solutions, but as an asset feeding disputes for political visibility within and across the state apparatus. This is particularly noticeable in São Paulo, where parties tend to establish ad hoc alliances and politicians act with a certain degree of autonomy, often voting against the instructions of their own parties.

If the Chamber mediates city noise, it is also true that noise mediates the Chamber. By closely following the legislative processes of genesis, revision, voting, and vetoing according to the requests and interests of residents, sound specialists, Evangelical leaders, lawmakers, mayors, and bar owners, it is possible to see how noise intersects with conflicting citizenship ideals. As in other parts of Latin America, noise legislation, and the very definition of noise,[62] is entangled in various projects of social organization at the interstices of differentiated and insurgent citizenships.

The MTST's protest noise in front of the Chamber is part of a more recent insurgent citizenship paradigm, in which noise becomes a political statement about the right to occupy, to be heard, and to demand inclusion. Living in a country of blatant differentiated citizenship, where laws are frequently deployed to keep the poor outside the rules of the game (outside the political circles), the protesters know it is not wise to wait for the Chamber to decide their future. They are acutely aware that to be part of the city *as citizens* they need to infiltrate the lawmaking process as much as possible. The initial vignette is thus a reminder that, just as powerful groups have attempted to use noise to reshuffle the city margins through urban planning and policing approaches, the margins have also found ways to deploy noise to reshuffle the center.

The analysis of Tripoli's noise ordinance demonstrates how Evangelical churches have penetrated lawmaking. Their interest in noise control relates to the ways in which their religious services are deeply enmeshed in intense devotional affect. While nonevangelicals often conceive of such sonic presence as crass and inconsiderate, Evangelical "noise" is also a strategy to expose nearby residents to the power of the Evangelical message. In the last twenty years, the organized action of religious lawmakers (who can move across party lines) throughout Brazil has affected not only noise control, but also broadcast licensing (Mariano 2004), women's and LGBT rights (Araújo 2016), and tax exemption.

---

[62] See Ochoa (2006, 2014).

Hato's 1:00 a.m. Law follows a premise popular since the 1990s, establishing a causal relationship between noise, alcohol consumption, violence, and crime. Although the law was criticized for its negative impact on the economy and imprecise wording, it prevailed thanks to the popular (and not entirely incorrect) perception of nightlife as a major source of violent activity in the city. When critics pointed out that the law did not close the bars for the reasons it was originally intended, it still had traction owing to its impact on nightlife noise. The ordinance's relatively heavy fines and the high costs of soundproofing have stimulated an increase in alternative social gatherings and the use of legal loopholes (e.g., having proxies as legal bar owners or tenants).

The last debate briefly discussed recent attempts to target noise traffic with the creation of a noise map. The very fact that traffic noise is virtually the background against which any other noise in the city is measured indicates its unique status. The Chamber has often ignored (and sheltered) this noise, and the few attempts to regulate it have been repelled.[63] The controversy around traffic noise is not limited to lawmakers, but common among sound specialists as well. As we saw in the previous chapter, the revision of Brazil's two most important technical standards for noise regulation, which most noise ordinances refer to, has been marked by disagreements over the inclusion of traffic noise.

In this chapter, we heard how different sounds resonate inside the São Paulo Municipal Chamber. By mounting a giant ear and bringing together lawmakers and the city residents, sound specialists attempted to pressure politicians to include cars, buses, and trucks in a new political circle. With the protesters' drumming, we heard the noise of two different political circles juxtaposed at the decisive moments when the lawmakers were voting on the Master Plan, a crucial piece of legislation that will change the distribution and association of actors in the city. Both the sound specialists inside the Chamber and the protesters outside framed the existing logic of the city as unsustainable because it excluded certain actors from the political circle—trucks and buses, homeless and jobless residents. The outside noise articulated by the protesters had some effect, as the approved Master Plan increased central areas for social land use.

The sound-politics of protest drums and eardrums, of small fleshy and giant inflatable ears, resonating inside and outside the Chamber, casting and recasting claims of belonging and existence, shaping governance through noise and noise through governance. Rather than a homogeneous project that revolves around access (to land, transportation, sewage, public health, etc.), noise-related citizenship

---

[63] See, for instance, Bill 853/1995 and Bill 707/1997. The former proposed noise limits based on urban zones, the former proposed fines for noisy bikers.

claims are highly heterogeneous. As we will continue to see throughout this book, this is partly because environmental noise is itself heterogeneous, difficult to measure, and ephemeral. Oscillating between taste, civility, and public health, it requires the constant monitoring of sound sources, which involves the challenging task of deciding which sounds can and should be controlled. It is not despite, but rather *because* environmental noise is such a constant, boundless, multifaceted, and affective entity that it has proven to be a fertile area of political negotiation in large urban centers like São Paulo.

> The State, like any regulatory apparatus, *follows* that which it regulates. Its applications are always retrospective, sniffing out and running after feral belongings it must attempt to recoup, to re-channel into State-friendly patterns.
> BRIAN MASSUMI, *Parables for the Virtual* (2002, 83)

4

ADMINISTRATIVE FLOWS

ONCE INTERESTED GROUPS enter the political circle and manage to convert issues into consensus and consensus into governmental action, some type of administrative reshuffling often follows. Creating and revising regulatory measures puts the government in two opposing paths. One path is expansive: the increase of state presence in everyday life through new and more sophisticated legislation. The other path is the minimalist: it is invested in the decrease in the amount and complexity of legal documents connecting state and collective life. Politicians get votes and bargaining power by inserting themselves into specific issues and moving the circle toward state expansion or contraction. What I am referring to here is the rather basic debate on the quantity and quality of documents and institutions that should be attached to the inhabitants of a given jurisdiction. Rather than turning an issue around to transform multiplicity into unity, we are dealing with the back and forth movement between legal documents and the series of actors necessary to do what these documents request. Only with this gesture can the state perform the desired convergence between the law and the reality it projects.

I call "administrative flows" the movement between governmental centers of execution and the population it oversees. In this movement, we see how the state is constantly "sniffing out and running after feral belongings it must attempt to recoup," as Massumi puts it. In the past chapters, we already have learned a few things about the executive branch. For instance, we saw that time and time again, it has been accused of perpetuating a *disjuncture* between law and law enforcement, with the proliferation of "dead laws"—documents that exist on paper but have no relation to the state's disciplinary performance. We also saw that noise control has often sat uncomfortably in the gap between the state and city administrative functions. While the police (arguably) have resources, territorial permeation, and a firmer hold on noise as public disorder, they are part of the state government. Municipal agencies, on the other hand, have fewer resources but more legal backing to effectively punish noisy individuals. Finally, we learned that the executive branch in São Paulo deals with the Chamber by turning issues into bargaining devices in multiple political circles.

In this chapter, we are going to focus on two law-enforcement institutions. The first is the police, one of the most visible state actors in the city and a recurrent choice for dealing with community noise. For reasons that will become clear throughout the chapter, the police have been at the center of both utopic and dystopic futures for the city (in both cases due to its coercive apparatus) and a constant source of political disagreements on the priorities and definitions of public security. The second institution is the Urban Silence Program (*Programa de Silêncio Urbano*, or PSIU), a municipal agency created in 1994 to enforce Tripoli's noise ordinance. Compared to the police, the PSIU is a smaller, more focused, and more specialized law enforcement agency. In following these two institutional circuitries, I suggest how encoding laws into administrative courses of action requires coordination between and within state institutions.

After tracing the administrative flows inside the police and the PSIU, I suggest a comparative analysis of the two institutions. By following noise complaints as they circulate within each of these governmental bodies, I argue that, different than the groups considered so far, the main concern for the executive branch is not so much ontological or epistemological issues surrounding noise, but rather consolidating (1) legal-bureaucratic stability, (2) budgetary principles, and (3) political party affiliations. Regarding the first factor, I draw on Michel Foucault to show the disjuncture of disciplinary and judiciary power mechanisms in both institutions. I relate these mechanisms to "nuisance" and "decibel" as the two main paradigms of noise control initiatives. Budgetary concerns relate to, among other things, conceptions of state performance and how public institutions increasingly face the double challenge of generating revenue for the city and not interfering with profitable activities in the private sector. I approach the third issue by contrasting the Worker's Party (PT) and the Brazilian Social Democracy Party (PSDB), the city's two most important parties in the last ten years. Although it can be challenging to classify both parties in terms of ideological commitment, I maintain that the

PT continues to operate around a version of welfare state whereas the PSDB has further embraced a neoliberal conception of state participation.

THE MILITARY POLICE

The phone rings. The police officer answers:
—Military police, help desk, how can I help you?
Staring at the screen, he clicks on "Offense Category." He selects and clicks on "Offense 1: Peace Disturbance."
—What is the address?
He enters the address into the system.
—What number?
He enters the house number into the system.
—What district is that? Do you have a point of reference?
He types "next to bakery 'Nova Orleans'" in the comments box.
—Okay. Where is the noise coming from?
. . .
—Ok. Yes. . . But . . . but you would have to go to the Police Department and file a repo . . .
. . .
—I don't know, ma'am. . . We just answer the calls. We don't control the dispatch. But I'm going to include a request to expedite this service, okay?
. . .
—You're welcome. Your request has been registered. Just wait to be assisted. You can always count on the military police. Goodnight.
They always finish a call with "You can always count on the military police." He disconnects the call and turns to me.
—Once a caller got angry with me bec . . .
The man seated next to him interrupts:
—Drunken guy with a two-year-old?
Noticing the question was too vague, the man elaborates:
—What category is that? Maltreatment? Abandonment?
—Yes, maltreatment.
—Yeah . . . Maltreatment of his liver!
We laugh. These jokes help us to lighten the mood and stay awake. He continues what he was saying:
—Once a caller got angry with me because I didn't have his address in the system. He said, "No record of streets in the slums, right? Nice job the

military police is doing, huh?" But I mean, what am I supposed to do? We get all types of calls and random complaints.

It is another Friday evening at the *Centro de Operações da Polícia Militar*, the Military Police Operations Center (COPOM). Throughout the night, the roughly ninety officers working in this large room will process thousands of requests and inquiries from callers across São Paulo state. When people in São Paulo dial 190 to get police help, it is to the officers in this room that they talk first. On weekends, the majority of the calls are related to community noise, a nonemergency case placed within the larger legal category of "peace disturbance." As we saw in the previous chapter, Article 42 from the Penal Contravention Laws (PCL) establishes as a misdemeanor (I) shouting or causing uproar, (II) conducting noisy or annoying activity in disagreement with the legal prescription, (III) using sonic instruments or acoustic signals abusively, or (IV) provoking or not preventing noise made by animals under one's guardianship.

In the corner of the room, six officers answer calls related to the Public Orientation Service (*Serviço de Orientação ao Público*, or SOP), responsible for nonemergency issues. After entering the noise complaint provided by the caller into the categories available on the police database (a first step in the gradual transformation of events into potential offenses), the call operator sends the information to the dispatch center, located in the next room. The dispatch center officer then forwards the request to the police battalion of the area. Once the information gets to the battalion, the commandant decides, depending on resources available, whether to send a patrol car to the noise scene. If the patrol officers take on the request, they will decide whether to engage with the incident—this will depend on the resources *they* have to cope with it. Every decision by every individual in this network is entered into the system in order to terminate the request. The information travels quickly and wirelessly, allowing officers to check each other's actions at any time. If the complainant dials 190 to complain about the same noise again (which usually happens), the call operator can see on his screen what decisions have been made.

If the commandant decides to send a patrol car, and *if* the officers decide to engage with the incident, the complainant is invited to go to the civil police department to file a police report. *If* the complainant decides to go, the issue will move from the military to the civil police. In Brazil, the military police (*Polícia Militar*) are responsible for ostensive policing whereas the civil police (*Polícia Civil*) are in charge of investigating and solving crimes with detective work and forensics. In that sense, the civil police are a necessary middle point between the military police disciplinary power and the judiciary punishing power. Public security in Brazil thus relies on the (not always smooth) coordination between the military and civil police. Civil police departments, headed by a local commissioner, process police reports and are located next to the military companies. The military and the civil police are each headed by a chief of police, who answer to the state's Department of Public Security.

At the police department, the noise complainant will describe the issue to the civil police officer. The police report may lead to a *police inquiry* (*inquérito policial*), conducted by the commissioner. Depending on the complexity of the case, the commissioner might summon witnesses, the alleged perpetrator, and the victim, before submitting the inquiry to the criminal judge. In this document, the commissioner explains either that there is no offense (in which case she will suggest the judge shelve the case) or say that an offense occurred—and the case should move forward. At this point, the noise complaint will leave the executive sphere and enter the judiciary branch through one of the channels to be examined in the next chapter.

Each of the more than thirty battalions active in São Paulo is headed by a military police officer (such as a lieutenant colonel) who oversees three or more police companies, one of which is the equivalent of a SWAT team. Companies are led by captains and oversee a territory within the limits of the battalion's area. Below the captain, the sergeant supervises the external operations of each company. Captains further divide their territory into sub-sectors according to its spatial extension, population, number of police reports, and other socioeconomic characteristics (such as the presence of slums). There are five types of police operations: radio patrol (which receives the dispatches from COPOM), school patrol, community police, special force, and motorcycle police force (*Ronda Ostensiva com Apoio de Motocicletas*, or ROCAM). Companies can manage the first three operations; only battalions can deploy the other two.

Every fifteen days, the captains across the city receive updated indicators to help them establish the time of day and sub-sectors in which the police need to patrol. The police include five types of crime in their statistical analysis to determine the allocation of resources: homicide, grand auto theft, car robbery, other types of theft, and other types of robberies. The police online network brings the data from the bulletin of occurrences of the police units, which are recorded in digital media and stored in a database. A piece of software then analyzes these data and creates a crime map, which identifies the highest crime points, separated by cities, neighborhoods and streets, as well as by days and times. With the data, it is possible to do Intelligent Policing Planning (PPI) and define the route of each patrol vehicle, according to the needs of the place.

While infractions usually turn into a police report in cases of theft or violent crime, they stop short of doing so in cases of noise complaints. Although the police continue to improve internal mechanisms for accurately classifying, mapping, and preventing public security problems, community noise not only continues to grow but insists on remaining unsolved. Instead of being eliminated from the streets, community noise is entering the police infrastructure and clogging their system. Environmental noise becomes administrative noise. COPOM's director, Major Carlos Tenório de Almeida (known inside COPOM as Major Tenório), whom I interviewed in 2012, explains the situation:

> We answer roughly 40,000 calls every day. On weekends, due to peace disturbance incidents, we have roughly 60,000 calls per day. Of every ten calls we receive on weekends, six are related to peace disturbances. We dispatch a police officer for two out of these six calls. The officer goes to the place, but usually *neither* incident turns into a police report. (Personal communication, June 2012)

The authority of statistics, which I discuss later in the chapter and which Major Tenório and his colleagues are so fond of, is a crucial element in what he defines as the military police's "scientific approach" to crime. According to this premise, the more you invest in classifying and analyzing the data that *enters* the institution through input devices such as the call center, the more effective the institution will be in coordinating action; also, the easier it will be to convince the governor and public security secretary that the police need more resources. Most of these noise complaints do not become police reports, the central data with which the military police operate. According to Major Tenório, only 20% of all calls do. If the problem is so prevalent, why are people not reporting it? Major Tenório explains:

> The person calls us on a Saturday at 11:00 p.m. to complain about the noise. This is not considered a police emergency, so we transfer it to SOP. SOP officers can either convince the complainant not to move on with the complaint (if it's something unreasonable) or dispatch a police car—because the complainant wants to go to the police. If he says he doesn't want to go to the police office, we say "you need to be patient then, because without your presence the commissioner is not going to do anything." So, we are going to send a police car whenever it's possible. Since the complainant doesn't want to go to police station to file a report, all we can do is to ask the noisemaker to turn the volume down.

Although they receive a large amount of peace disturbance complaints, these are not taken into account when determining the allocation of police resources precisely because they do not generate reports. Although the São Paulo State Military Police have recently tried to create alternative channels to unclog the system, complainants still prefer to avoid filing a report for fear of being exposed or because of distrust in the administrative process. Dealing with noise without eliminating it risks damaging public opinion toward the police, and some officers told me that tackling this problem was a major obstacle for the police in gaining respect in communities. The clogging of the police administrative flows caused by noise complaints has become so serious that several police officers have tried to offer alternatives.

One strategy to unclog COPOM's backup was to create an online complaint system. The system allows users to specify the space where the noise is coming from and the type of noise. The list includes car alarms and rowdy children in the street; construction and parties in residential spaces; and gas stations, churches, and bars in commercial spaces. Users need to type in personal identification numbers and choose whether they are going to contact the police officers once they get to the location and whether they would be willing to go to the police station to file a police report (see Figure 4.1).

This might help the institution, but it is not exactly what Major Tenório has in mind. His research project focused on precisely this issue.[1] In the introduction of his research project he writes:

> In this project, I contend that administrative punishment is the primary resource for solving this problem [of peace disturbance], which plagues so many people. The police have this instrument in situations of noise infractions related to traffic. In these circumstances, the police officer does not depend on other public authorities—he can take action and solve the problem himself. However, with other noise sources, the police cease to be the ultimate decision maker and solving the problem becomes more difficult.

For Tenório, the goal is to make community noise an administrative infraction under the jurisdiction of the military police, exactly like traffic infractions. But to become the "ultimate decisionmaker," the military police would have to buy hundreds of sound level meters and instruct police officers on how to take measurements properly. This would require not only budgetary reshuffling but a change of perspective and priorities. What exactly is public security? Other police

---

[1] Majors need to submit a research project to be promoted.

FIGURE 4.1. In 2013, the military police started to offer an online noise complaint channel for noise complaints. The complaint form asks complainants to identify the origin of the disturbance: street, residence, or commercial venue. The "street" option (pictured here) includes pancadão/street party, sound equipment, alarm, construction, vehicles in the street, agglomeration of people, and children. The "commercial venue" option includes gas station, church/worship space, nightclub, bakery, bar, restaurant, and car wash. The online complaint form also asks the complainant, "Will you contact the police car?," "Are you going to the police station to file a complaint?" If the user answers "no" to any of these questions, a note appears explaining that this will make it more difficult to hold accountable the noisemakers.
Source: http://www2.policiamilitar.sp.gov.br/ocorrenciaweb.

officers counter Tenório's viewpoint by asking why they should invest more time and money with peace disturbance if the military police cannot handle serious crime, such as homicide and theft. Is noise even a police matter?

For a group that includes Major Tenório, the politicians defending the 1:00 a.m. Law in the previous chapter, and other actors that we will meet in the following chapters, there is indeed a *strong* link between peace disturbance and crime. If the military police are constitutionally obliged to work as a preventive force against crime, then clearly peace disturbance should be a priority. For others, however, this link is *weak* and a mere diversion from life-and-death issues. Punishing noisemakers is a task for the city administration, this group argues. In fact, most of São Paulo city-level traffic infractions, to use Major Tenório's example, has been under the Traffic Engineering Agency's jurisdiction (*Companhia de Engenharia de Tráfego*, or CET) for more than thirty years. One thing one can hardly disagree with Major Tenório about is the general expectation that the military police can solve problems like no other institution. By considering the impressive number of noise complaints the COPOM receives every single day, it is safe to say that residents do hope the police can eliminate community noise. This is not only due to the institution's resources and "scientific approach." It also relates to its long history of "coercive methods."

Thomas Holloway explains that in the colonial period, despite attempts to emulate England, France, and the United States, the police in Brazil grew from indigenous practices and assemblages. As Holloway explains, "The hostility between the forces of repression and the sources of resistance in Brazil is related to the imposition of apparently modern bureaucratic institutions of control on a society that was lacking in other fundamental attributes of modernism [such as equality before the law]" (Holloway 1993, 6). The Military Division of the Police Real Guard, created in Rio de Janeiro in 1809 at a time of full-fledged slavery and a fearful elite, was responsible for preventing nightly gatherings and making sure stores, cafes, and bars closed by 8:00 p.m. (Faria 2007, 57). Members of slaveowner families who decided to take the streets to celebrate, get drunk, and provoke disorder were spared from sleeping in the city's miserable detention houses. In his 1890 novel about everyday life in a Rio de Janeiro tenement, author Aluizio de Azevedo writes: "With the excuse of preventing or punishing gambling and drinking, the urban guards invaded rooms, broke everything, put everything in shambles. It was a matter of long-standing hatred" (quoted in Holloway 1993, xix).

Holloway shows that due to this disjuncture between a modern institution of social control and the country's pervasive differentiated citizenship, the police invested considerable time and energy with "offenses against public order" such as vagrancy, verbal insult, and public drunkenness. In that way, "The seamless association of behavior that is nearly universally condemned (theft or murder) with the victimless violation of arbitrary rules (curfew violation) or symbolic defiance of authority (disrespect) is one of the slickest examples in the history of class society of imposing guilt by association and evil by extension" (Holloway 1993, 9). André Rosemberg suggests that historical approaches to the police tend to follow the instrumentalist perspective, which sees the institution as the local elite's disciplinary puppet; the self-narrative of a glorious past populated by heroes; and the liberal perspective, which takes the police as modernizing force of a rational and democratic state (Rosemberg 2010; Bretas and Rosemberg 2013). In his analysis of São Paulo Permanent Police Corps (*Corpo Policial Permanente*, which preceded the São Paulo Military Police) in the 1870s and 1880s, Rosemberg shows that most police officers working in the city's disorderly streets came from these same streets, as recruitment did not depend on race, class, or literacy (some were fugitive slaves using fake names). Whereas European immigrants were the choice for most professions in the province, poor Brazilians saw in the police a path to citizenship rights and social ascension. In contrast to the Permanent Police Corps, the Urban Guard (*Companhia de Urbanos*), created in 1875, was a demilitarized preventive police force (literacy was mandatory for entrance) with the mission of modernizing what was then a relatively small city of 30,000 inhabitants. Following the European and U.S. police forces, the Urban Guard started to divide the policed area into zones to be continuously monitored. The city space was classified according to a range of socioeconomic niches; the "dangerous class" was properly identified and kept "eternally under suspicion" (Rosemberg 2010, 268).

With the end of the Brazilian Empire (1899), the new Republican Federalist government gave each state autonomy to administer the newly created Military Police Corps, which was part of the Public Force (*Força Pública*) together with the Fire Corps. In the early 1910s, concern with national security stimulated a change to allow the army to deploy the police force if necessary. This made the police officially part of the Army Reserve and stimulated the use of a similar military ranking system in both institutions.[2] In 1926, the state created the Civil Guard as a demilitarized auxiliary agency to the Public Force.[3] Until the 1960s, the Public Force was accountable for safeguarding public property, with the Civil Guard handling everyday policing. In 1970, the Civil Guard and the Public Force merged to officially become the São Paulo State Military Police. In 1986, mayor Jânio Quadros (whom we already met in Chapter 1) created the Metropolitan Civil Guard to protect "municipal goods, services, and facilities, and to collaborate with public security" (Law 10115/1986).

The 1967 federal constitution designated the military police as the institution responsible for maintaining "internal order and security." During this period, at the height of political repression after a series of presidential decrees virtually removed basic civil and political rights, the military police established "scientific" torture methods against political prisoners (Arns 2001, 32). In the car-manufacturing sectors of greater São Paulo, the civil police coopted employers to identify and neutralize political dissidence in the workforce. As mentioned in the previous chapter, it was from this moment of systematic violence and abuse that a group of politicians, scholars, religious leaders, and factory workers created the Worker's Party in the late 1970s. In the streets, the highly armed military police special forces (similar to the SWAT teams in the United States) conducted a series

---

[2] The military police are organized into two groups: police soldiers and police officials. The former has this ascending hierarchy: soldier, corporal, third sergeant, second sergeant, first sergeant, and subtenant. The latter, which requires further specialized training, includes (in ascending order): second tenant, first tenant, captain, major, lieutenant-colonel, and colonel.

[3] The Civil Guard appears in the second anti-noise wave discussed in Chapter 1.

of operations in the city with random checkpoints, roadblocks, and background checks (Fernandes 1989).

In the 1970s, public security officials articulated a straightforward rationale of police violence, what Heloisa Fernandes describes as the "discourse of suspicion" (Fernandes 1989, 125). As innocent citizens continued to die in police operations, the military police chief and the public security secretary came to the press to frame the situation within an excess/omission binary. "Only by trying to make it right one can make mistakes. The negligent usually doesn't make mistakes!" explained the Public Security Secretary in 1976 (quoted in Fernandes 1989). With the support from the press, these officials justified violence abuse by the military police, such as the case of three teenagers who ended up chased and shot by the police (with two submachine guns) for not pulling over, as a byproduct of the police officer's commitment to protecting the "good citizens." Not all state agencies agreed with this approach, however. Echoing the view that the police were permeated by "uncivilized" individuals prone to unjustified violence, a representative of the public prosecutor's office at the time criticized the institution's recruitment mechanisms and the "arbitrary arrogance and cruelty that arises when men of lower education are invested with authority" (quoted in Fernandes 1989, 126).

After the redemocratization, the 1988 Constitution replaced "national security" with "public security." Although the new document made important changes such as moving incarceration decision to the judiciary, it preserved most of the existent structure—a militarized state-based police force loosely attached to a civil investigative institution. Some scholars argue that the fall of the military regime did not change the police modus operandi of protecting the state rather than the citizens (Lima et al. 2016). On the one hand, social conflicts—regardless of their nature and context—continued to be framed solely within the criminal sphere. On the other, certain social groups (black men in particular) continued to be seen "as intrinsically dangerous and an object of constant surveillance and neutralization" (Lima et al. 2016, 57).

Fear of crime increased when Brazil transitioned back to democracy, which led to support of illegal and undemocratic responses by the state, a process Teresa Caldeira and James Holston describe as paradoxical (Caldeira and Holston 1999). A 1999 survey conducted by the University of São Paulo showed that 61% of Paulistanos agreed with the statement "it is difficult to feel protected by the laws" (Cardia 1999, 55). Police abuse continued in the post-dictatorship period. In 1992, following a riot in Carandiru (then the largest prison in Latin America), the São Paulo Police killed 102 unarmed prisoners after they had already been surrounded—in 2016, a court annulled the sentences of 74 police officers for the massacre. In 1997, Brazil's main TV network broadcasted footage of police officers engaging in brutal violent acts (including one execution) against residents of a favela in greater São Paulo. In the late 2000s São Paulo, "more people [were] killed by police every two years than the military dictatorship killed during its entire twenty-year reign" (Arias and Goldstein 2010, 2).

In 1997, in response to an already shaky relationship with civil society, the São Paulo military police adopted community policy strategies, "an amalgam of previous Brazilian practices and the 'original American concept' of community policing" (Ferragi 2010, 33). The São Paulo police trained thousands of officers in community policing and installed community-based sub-stations across the city. A central rationale in American community policing has been the "broken windows" approach, briefly considered in the previous chapter. Put forward by James Wilson and George Kelling in an influential 1982 article, the theory maintains that there is a strong relationship between crime and the environment in which it takes place. As the authors explain, "One unrepaired broken window is a signal that no one cares, and so breaking more windows costs nothing" (Wilson and Kelling 1982). Community policing stimulates a closer relationship between residents and police officers, and stimulates the discussion of a wide range of neighborhood problems (beyond strictly public security issues). It also gives police officers a more direct environmental grasp of social interactions and facilitates their perception of disorder as an avenue to crime. In both cases, residents and police officers frame community noise as a broken window that needs to be identified and fixed right away to prevent the occurrence of more serious problems.

The Community Security Councils (the CONSEGs, to which I return in Chapter 6) provide another link between noise and security in São Paulo. The councils, established in the state of São Paulo in 1985, encourage the cooperation between the police and civil society. In the city of São Paulo, these meetings take place monthly and are organized by district. Each council is presided by a citizen (who must live or work in the area) and attended by the area commissioner (civil police) and captain (military police). Officials from the subprefecture, municipal transit agency, and the Metropolitan Civil Guard are also usually present to answer questions.

Some police officers told me in private conversations that CONSEG participants often include more serious acts in their descriptions of environmental complaints (a dark park or a loud bar) in order to persuade the police to take action. Captain David, who works in the Jabaquara district, explained that residents often talk about "loiterers *with some type of weapon* or teenagers *using some type of drug*. They most certainly can't tell if someone is armed just by looking from a distance, from their windows. But if it is just a guy hanging out in the plaza, we won't go because there is nothing wrong with that" (personal communication, May 2012). As a result, what could be seen as simply dirt, lack of public lighting, or loud music, is gradually linked to "suspicious" activities. Similar to what Benjamin Chesluk observed in his ethnography of security council meetings in New York City in the early 2000s, citizens request police intervention by learning how to describe their problems from the police's perspective, what the author calls "broken windows stories" (Chesluk 2004, 254). These points of exchange between the police and São Paulo residents are part of a continuum that establishes what David Matless calls the "moral vocabulary of a landscape": "a language for harmonious human-environment relations" (Matless 1995, 88).

We see thus an interesting dynamic taking place with public security and sound-politics in São Paulo. In criminalizing space and spatializing crime, purifying the institution and stimulating the purge and *shared* surveillance of disorderly acts, the police use their disciplinary power to maneuver the public opinion and circumvent legal obstacles preventing them from quieting noisy people. I have retraced the trajectory of the police in Brazil to highlight that granting this institution authority to tackle noise-as-disorder is not exactly a new debate. What are the risks of turning the police into a disciplinary device that operates by establishing guilt by association and evil by extension? It is because, since the colonial period, they are known as a professional group that is willing to follow or "bend" the law to enforce quiet that police officers enter the sound-politics debate in São Paulo in a unique fashion. Still, some police officers seem to resent the concern of human rights groups that the state should be very careful on giving them legal authority to act due to a history of violence and abuse. As it will become clearer in Chapter 6, when we examine the spread of street parties in the city's poor suburbs, the heterogeneity of issues under the police's belt allows leaders such as Major Tenório to more easily enlist other state agencies, as virtually any public problem can be engulfed in the "broken windows" "quality of life" paradigm.

The PSIU

At PSIU's front desk, a man in his forties holding a binder walks in:

> MAN: *Hi, I was reported. Here is the letter* [see Figure 4.2]. *I want to follow what the law requires. The important thing is not to have my church closed.*
> PSIU OFFICIAL: *Did you bring your documents?*
> MAN: *Yes, I have everything. I turn off the microphone at 8:30 p.m., to avoid trouble. One day I finished the services at 9:30 p.m., and they came here to report me. I think the best thing is to remove the drum kit and leave the microphone on*

PREFEITURA DA CIDADE DE
SÃO PAULO

São Paulo, 08 de 2 de 2012.

COMUNICADO 2012

Sr.(a) Responsável,

A Divisão Técnica de Fiscalização do Silêncio Urbano, órgão público competente pela coordenação do controle e fiscalização de atividades geradoras de poluição sonora, bem como de bares com horários de funcionamento irregulares e demais eventos perturbadores do silêncio urbano, em desrespeito às Leis Municipais nº 11.501/94, nº 11.986/96, nº 11.938/95 e nº 12.879/99, e legislação pertinente, vem, por meio deste COMUNICADO, informar que a Prefeitura do Município de São Paulo recebeu uma reclamação/denúncia, relativa ao estabelecimento.

Esta Divisão Técnica esclarece, por oportuno, que o Programa do Silêncio Urbano - PSIU tem por objetivo prioritário conscientizar o público leigo quanto aos efeitos prejudiciais causados à saúde, pelos ruídos excessivos. Assim, solicita a sua compreensão e estreita colaboração com este Poder Público, visando a preservar a saúde e o bem estar da população que trabalha, freqüenta, reside ou transita nas proximidades de seu estabelecimento.

Desnecessário ressaltar, outrossim, que algumas providências elementares já podem ser imediatamente adotadas a fim de evitar novas reclamações/denúncias, como, por exemplo, o respeito ao horário de trabalho, a redução das fontes de ruído, o controle do nível de qualquer fonte sonora com transmissão ao vivo ou sistema de amplificação, o fechamento de aberturas da edificação, a instrução adequada dos empregados na utilização de fontes sonoras e produção de ruídos, dentre outras adequações.

Para maiores informações, favor contatar esta Divisão Técnica, sito a Rua Líbero Badaró, 425 – 36° andar – Centro, no horário de atendimento ao público, das 10:00 h às 16:00 h, as segundas, quartas e sextas-feiras.

No aguardo das providências indispensáveis, esta Divisão Técnica recomenda o estrito cumprimento à legislação vigente, no prazo máximo de 30 (trinta) dias corridos, contados do recebimento deste COMUNICADO, pois seu não atendimento acarretará, após a constatação da irregularidade, e por dever de ofício, a aplicação das penalidades previstas nas leis supra citadas.

Agradecendo antecipadamente a sua colaboração,
Divisão Técnica de Fiscalização do Silêncio Urbano

Estabelecimento ou local de trabalho objeto da reclamação:

Denominação:
Endereço:

Complemento:

Subprefeitura: « »

São Paulo – SP.

FIGURE 4.2. PSIU's notification letter from 2012. After referring to the city's noise ordinances, it reads: "This Technical Division clarifies that the Urban Silence Program—PSIU—has as its main objective to make the lay public aware of the harmful effects caused to health by excessive noise. Therefore, it requests your understanding and close collaboration with this Public Power, in order to preserve the health and well-being of the population that works, frequents, resides, or transits in the vicinity of your establishment.

"It is unnecessary to point out that some elementary measures can be adopted immediately in order to avoid new complaints, such as respect for working hours, reduction of noise sources, control of the level of any sound source with live transmission or amplification system, closing of building openings, proper instruction of employees in the use of sound sources and production of noise, among others.

[. . .]

"While awaiting the necessary measures, this Technical Division recommends strict compliance with current legislation, within a maximum period of 30 days, as of the receipt of this letter, since its noncompliance will entail, after verification of the irregularity, and as official duty, the application of the penalties fixed in the abovementioned laws."

at a low volume. The musician at my church is already mad at me because I don't let him play longer. But I'd rather lose him than have to close my church.

The PSIU official opens the process in the system: *You have a Pentecostal Church. The complaint was made on November 24. It says your church was making noise from 7:00 a.m. to 9:30 p.m.*

MAN: That's a lie! I don't open the church in the mornings. They really want to break me, huh? I don't offer services during the day. Now that I know about you here, I'm going to report them too! This man who reported me, the other women next to me, they throw parties that last the whole night. Can I report them too?

PSIU OFFICIAL: Yes, you can call 156. But you should worry about your problem now.

MAN: . . . The whole night!

PSIU OFFICER: If the PSIU goes there, you may get a fine, and then it's going to get complicated for you. Let's sign in the Acknowledgment Form. I'm going to put that in the system. If we receive another complaint, we are going to take action . . .

MAN: I'm going to remove the drum kit. But we pay taxes; I can't leave my microphone on even before 8:30 p.m.?

PSIU OFFICER: There's no period in which you can make noise.

MAN: But everybody makes noise! Just because I have my church I can't work?

PSIU OFFICER: You can work. But there's a complaint about your church. You should worry about your church.

MAN: Yeah, I better remove that drum kit. The Baptists don't play loud music . . . neither do the Catholics. We don't have a lawyer, [so] we can't squabble with the entire neighborhood. I need to turn the volume down, play the music inside the house only. I know they are dying of jealousy because I have a job, because I bought two motorcycles. After I bought those motorcycles it became hell to live on that street. But I'm scared of being fined.

The man signs the Acknowledgment Form and leaves.

As we saw in the previous chapter, in 1994 a new comprehensive noise ordinance for the city of São Paulo came into effect. The ordinance (updated in 1996), known as the "Noise Law," was created in response to noise complaints against bars, restaurants, gas stations, churches, and construction sites. The ordinance establishes a fine of US$12,000,[4] to commercial venues without a proper license. According to the law, a venue that continues to operate without a license after the first fine receives a second fine (plus one-third of the first fine). A third inspection of

---

[4] See Chapter 3, Footnote 17.

the place leads to administrative closure. Venues with a valid license making noise above the limits established by the zoning law receive a fine between US$2,000 and US$8,000 (depending on the capacity of the space), also followed by a second fine and administrative closure. In both cases, the establishment that breaks the administrative seal is committing a crime and is thus liable for punishment.

The 1994 Noise Law (later incorporated by the 2016 Master Plan), was created to address the city's fragmented laws and ineffective law enforcement. The ordinance delegated to the municipal government the responsibility for creating an anti-noise agency equipped with the "mechanisms to manage complaints and [. . .] inspect and measure noise levels."[5] In October 1994, São Paulo mayor Paulo Maluf issued a decree creating the PSIU[6] (*Programa de Silêncio Urbano*, or Urban Silence Program), an agency dedicated to noise inspections and fines. It was agreed that the agency would operate across municipal departments, one of which would oversee its activities. The mayor gave the coordinating role to the city's newly created Environmental Agency, which would register the complaints and inspect the venues.

Already in its first years of existence, the PSIU's trajectory started to shift. In 1996, Maluf issued another decree moving the coordinating role to the Department of Food Supply, arguing that the "[the department] already inspects the hygiene of the kitchens in commercial establishments."[7] This shift reduced the number of inspection agents and strengthened the links between the PSIU and the service sector (i.e., bars and restaurants). According to Regina Sobral, who was involved in the creation of the Noise Law in the early 1990s, Maluf's decision was an attempt to avoid problems with major public construction projects, the mayor's main political asset (personal communication, July 2013).

In Chapter 3 we also examined the genesis of Hato's "1:00 a.m. Law." The ordinance prohibits bars from staying open after 1:00 a.m. without proper sound-proofing. Offenders receive a fine of US$12,000, with administrative closure in case of repeat offenses. In 2001, mayor Marta Suplicy, from the center-left PT, issued a decree assigning the 1:00 a.m. Law to PSIU. In 2003, she decentralized the Department of Food Supply—including the PSIU—by transferring them to the city's newly created thirty-one subprefectures.[8] According to Suplicy's decree, the shift was necessary to "standardize and unify the criteria, methods, and procedures concerning the control, licensing, monitoring, and inspection of

---

[5] Law 11501/1994.
[6] Decree 34569/1994.
[7] Decree 35919/1996.
[8] Since 2013, São Paulo has thirty-two subprefectures.

activities that generate noise pollution in São Paulo."⁹ With this change, the PSIU moved to the Department of Subprefectures. Debora Castelani, who worked at the PSIU between 1997 and 2014, describes her job during the decentralization period:

> When it was decentralized, one subprefecture coordinated the actions of an entire region. We received the complaints from the subprefectures in our regional sector and organized the inspections for each. The director, who stayed in an office downtown, verbally named a sub-director for each sector. It was a bad idea to decentralize PSIU because we ended up spread out and isolated in tiny rooms. We were expected to schedule the places to inspect, make the inspections, and process the technical reports... We had to do everything! And we could only do the inspections *if* we had a vehicle available, which depended on all the other activities of each subprefecture. (Personal communication, October 2012)

In 2005, when the city administration shifted from the PT to the center-right PSDB, the elected mayor appointed Moacir Rosado, known as the "Major" due to his police background, to direct the PSIU. Following Major Rosado's request, the mayor issued a decree re-centralizing the PSIU, arguing that "decentralizing the control and inspection of the activities that generate noise pollution [...] did not achieve the expected results."¹⁰ As Major Rosado explains:

> This system of creating regional sectors in the subprefectures created a mess. Who is responsible for doing what? Who has the authority? I took charge as the director in the midst of this whirlwind of responsibilities. I took on the job saying: "Look, I follow the principle of legality. I'm the director, so I'm in charge here." So we had to restructure the PSIU. I established the headquarters downtown and brought everybody in. We found out that each person had a different interpretation of the law. I told the mayor that I would need six to nine months to make the agency functional. (Personal communication, July 2012)

With privileged access to the mayor's office, Major Rosado was able to secure a large office area in downtown São Paulo, buy new equipment, hire new people, train the staff, and have dedicated vehicles for the inspections. Additionally,

---

⁹ Decree 43799/2003. Although Mayor João Doria has changed the name "subprefecture" to "regional prefecture" in 2017, I use "subprefecture" throughout the book.
¹⁰ Decree 45729/2005.

rather than responding to complaints, he used his connections with the police to expand the agency's operations, organizing large inspection blitzes in neighborhoods with a high incidence of noise complaints. As infractors continued to ignore the agency's actions, Major Rosado articulated a heavier reading of the law (see Figure 4.3):

> I researched the legal definition of "administrative sealing," and it includes putting large stone blocks in front of the establishment to keep the owner from opening it. The block is city property, so people are not allowed to remove it. If you remove the block, you are damaging public property, and then you are in trouble. We also started to wall up the entrance. The lawyers attacked me saying that was illegal. When the judge stated that it was legal, *then* everybody started to come to talk with me nicely. (Personal communication, July 2012)

Rosado explains: "During all my years with the police I worked on the street. I used to get frustrated when the bar was making noise but had all the paperwork in order. What could I do as a military police officer? So I had to just leave. This would often give the impression that we had made a deal with the

FIGURE 4.3. Rosado's "stone-block" interpretation of administrative closure. The image shows several concrete blocks labeled "Interdicted: City of São Paulo, Pinheiros Subprefecture" restricting access to a bar in the Vila Olímpia upscale district.
Credit: *Folha de São Paulo* (February 22, 2005), C3.

bar owner."[11] Since inspections under the 1:00 a.m. Law are relatively simple, requiring the inspector to check whether the bar is open and taking punitive administrative measures, Rosado started to move from passive to active noise control, supervising large inspections in entire districts with the help of the military police. Rosado is yet another proponent of the noise-crime link usual among zero-tolerance advocates:

> Bars open after 1:00 a.m. without acoustic insulation and security personnel are suspicious. I tried to change that law to 11:00 p.m., because when you close a bar at 1:00 a.m. you are throwing patrons on the street but not providing any public transportation. So you can create another problem. We put together these operations in a peripheral district for a few months and found a fair decrease in the crime rate in that region. Our biggest surprise was that aggression against women almost disappeared. (Personal communication, July 2012)

Rosado left the PSIU in 2007, a few months after the mayor stepped down to run for governor. In 2012, when I conducted fieldwork at the agency, many of Rosado's rearrangements remained in place. It was still centralized in a large office space in downtown São Paulo, but the staff had shrunk, and the large inspection blitzes were discontinued. The director was Wanderley Pereira, a retired police officer known inside the agency as "the Colonel" (the new mayor followed his predecessor in placing someone with a police background to direct the agency). The Colonel explained that the PSIU was receiving about two thousand complaints every month; 60% of complaints were related to leisure venues (bars, restaurants, and nightclubs). Most of the other complaints had to do with religious services and construction sites.

Unlike Major Tenório and his allies, the Colonel does not think the military police should deal with noise disturbances. "This is not what the military police would like to be doing—I know that because I'm a retired colonel. They have other priorities. Noise is an administrative issue" (personal communication, January 2012). For the Colonel, complaints that require prompt response fit under the peace disturbance situation, which the police handle straightway. "Here at the PSIU the demand is too great to respond promptly, which is why we need to schedule inspections. Besides, since we depend on the presence of police or metropolitan civil guard officers, we need to send an official request a week before so they can allocate the officers who will accompany the PSIU agents" (personal communication, January 2012).

---

[11] As we can see, the frustration showed by Major Tenório is not particularly new among military police officers.

The PSIU is part of the Department of Subprefectures, under the General Supervision of Land Use and Occupation. Besides housing the PSIU and serving as a mediator between the city and its thirty-two subprefectures, the department also enforces the Clean City Law and manages the Delegated Operation (*Operação Delegada*). The 2006 Clean City Law[12] was proposed by the executive branch to regulate public advertising posters and prohibit billboards in the city. By diminishing the visual pollution of advertisements, the law intends to highlight the architectural integrity of the buildings and increase the safety of pedestrians and drivers. The fine for the first infraction is around US$2,770. The Delegated Operation, which started in 2009, is an agreement between the São Paulo city and the São Paulo state that allows military police officers to conduct inspections as city officials in their off-duty time. The initial objective of this agreement was to fight pockets of unregulated street commerce in the central districts. As a report from the Department of Subprefectures explains, "At the same time that they are working in the Delegated Operation [. . .], these officers exert their police power in favor of public safety" (Prefeitura de São Paulo 2012, 23).

In order to better understand how the PSIU deals with noise in the city, we need to follow the agency's bureaucratic circuitries. The PSIU headquarters is divided into five main sections. The front desk orients people and answers phone calls from complainants wanting to know where their cases stand. As suggested in the exchange with the church leader at the beginning of this section, the front desk official also instructs people who have received a notification that a complaint has been made against them. They are asked to come to the PSIU office to sign an acknowledgment form.

The three other rooms are the director's office, the juridical sector, and the fine registration sector. The director is the agency's administrative hub, the gatekeeper who oversees and approves the input and output of documents (soliciting the military police to provide support for the inspections, responding inquiries from the Public Prosecutor's Office, approving inspections, etc.). The juridical sector deals with those who challenge the agency's actions in the administrative or judicial spheres. In short, they protect the PSIU against legal attacks and advise the director on existing and potential litigation. The fine sector is where fines are properly cataloged and become part of a case folder, which the staff enters into the system. Anything that happens related to the case (a new fine, a legal response from the bar, a report from the acoustician stating the bar now conforms to the law) is added to the folder. This sector archives the original copies of the fines. One staffer who moved to this sector recently told me that, in the few months she

---

[12] Law 14223/2006.

had been working there, the archives were broken into three times. "The fine remains in the system, so it doesn't affect the legal procedure" the staffer explains, "But maybe the burglar would take the physical originals to bribe the bar owners, as if that was the only version of the document. . . ." (personal communication, June 2017).

A large central area with partitions includes human resources, the ombudsman, the scheduling coordinator, and the programmers (see Figure 4.4). Each programmer is responsible for one of the city's five regional sectors (Center, West, East, North, and South). For instance, Ms. Teodoro is the programmer responsible for the East sector, which comprises roughly four million people in thirty-three districts. Early in the week, Teodoro updates the system with data about recent inspections. Each new activity must be carefully linked to past actions to keep the administrative flow as smooth as possible so that the inspectors know what has been done and what needs to be done—levy a fine, implement an administrative closure, and so on. After updating the system, Teodoro then puts together a list of inspections to be conducted later that week. Once the programmers finish compiling the lists, the scheduling coordinator organizes the inspections into

FIGURE 4.4. Monday morning at the PSIU headquarters: programmers enter into the system the actions conducted by the inspectors over the weekend.
Photo by the author, 2012.

blocks of around ten, puts the lists in sealed envelopes, and determines which inspection agents will be responsible for each list.

To examine the flow inside the PSIU's administrative circuitry, I propose we follow the dispute between Ms. Freire and Bar da Esquina. Ms. Freire is fed up with Bar da Esquina playing loud music every weekend night. The music rattles her bedroom window and deprives her of quality sleep, which she considers a right after a week of hard work. Unwilling to talk with the bar owner and tired of having the police come and "do nothing," Ms. Freire decides to seek help from the PSIU. She accesses the Municipal Public System online and fills out the complaint form (see Figure 4.5).

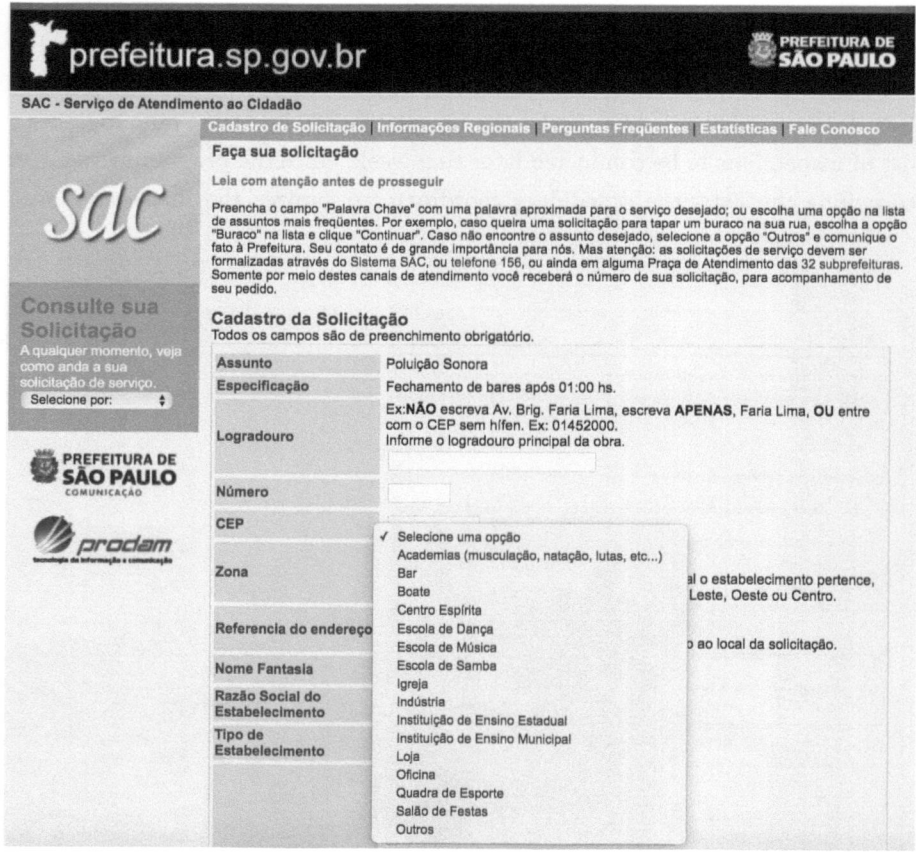

FIGURE 4.5. The São Paulo Municipal Public System website allows São Paulo residents to make noise complaints to the PSIU. In "specification" the complainant is asked to indicate whether the noise fits into the Noise Law or the 1:00 a.m. Law (as explained in Chapter 3, this was changed in 2016, when both ordinances were incorporated by the new Master Plan Law). In "type of venue" (pictured above) the user can choose between gym, bar, nightclub, spiritist center, dance school, music school, samba school, church, factory, state public school, municipal public school, store, vehicle workshop, sports court, ballroom, and others.

The PSIU programmer accesses Ms. Freire's complaint. The programmer includes Bar da Esquina in the list of places to be inspected on the following weekend at the time indicated by Ms. Freire in the form. Again, it is important to keep the administrative flow smooth: if the agents arrive either too early or too late, that can compromise their technical report—Bar da Esquina might claim it was not playing loud music at the time indicated by the complainer, or Ms. Freire might argue that the agents did not find any wrongdoing because they failed to arrive at the right time. The scheduling coordinator then assembles the inspection lists, based on the complaints coming from citizens (such as Ms. Freire's), from the Public Prosecutor's Office (usually a priority, as that involves cases that have moved to the litigious judicial sphere), and from the military police (for large operations such as the one we will see in Chapter 6). She then sends requirements to the Metropolitan Civil Guard or military police, which will provide security backup to the agents.

The scheduling coordinator randomly selects a crew of two inspection agents and a driver for each of the ten lists—one for each ordinance in the city's five regions. The scheduling coordinator places the envelopes for Saturday's inspections in the cabinet. One pile of envelopes is for the four engineers who deal with the "Noise Law," the other pile is for the "1:00 a.m. Law" inspectors. One officer, the liaison, arrives before the day crew leaves and after the night crew arrives (almost all inspections occur during nighttime). As a link between day and night activities, this person makes sure important and updated information is exchanged effectively between the two groups. He delivers the sealed envelopes to the agents when they arrive and makes sure the designated drivers get there in time.

The agents arrive at the PSIU around 7:00 p.m. They receive the envelopes and examine the addresses to figure out the best routes to optimize time in traffic-intense São Paulo. Unlike the 1:00 a.m. Law agents, the Noise Law agents (who will inspect the Bar da Esquina) make measurements with a sound level meter (SLM). Again, every point of passage separates a smooth from a bumpy flow: if an agent forgets to bring enough forms (which I observed once during fieldwork) or to check the battery of the SLM, the inspection will be cut short and the administrative flow compromised. After thirty-five minutes, the crew arrives at the Metropolitan Civil Guard station, where they meet the guards who will accompany them. The first three places on the list are closed. The fourth place is a bar located on a busy avenue. Although the sound level meter marks 72 dB (A), above the zone limits, the environmental noise on the street is too close to that value, preventing the agent from issuing a fine.[13]

---

[13] Recall from the previous chapters that in NBR 10151, on which the Noise Law draws, traffic noise is often the point of reference for evaluating a specific noise source.

They finally advance toward the Bar da Esquina. A few blocks before arriving, the agent measures the environmental noise: 55 dB (A)—this value will be the point of reference to establish how much noise the bar is projecting into the public space. At the Bar da Esquina, the agent measures the sound two meters away from the façade and 1.2 meters away from the ground: 72 dB (A). Not only is the value above zoning law limits, but the difference between environmental (55 dB [A]) and specific noise is wide enough. The agents go inside and ask for the bar license. Everything is in order. The agent then explains that the Bar da Esquina is making noise above the limits, which is 45 dB for a mixed zone of medium-to-high density between 10:00 p.m. and 7:00 a.m. the following exchange ensues:

> BAR MANAGER: *Sir, how am I supposed to know when my bar is being too loud?*
>
> PSIU AGENT: *You can't have this type of speakers with the windows of your bar open. They are projecting sound to the streets. That's why you got a complaint. You need acoustic treatment here.*
>
> BAR MANAGER: *Last week a PSIU agent came here and he told me the limit was 65 dB. So how am I supposed to know?*
>
> PSIU AGENT: *No, it's 45 dB after 10:00 p.m. If you didn't have the license the fine would be US$12,000. Since you have the license, it's going to be less. How many people can you fit in here? 100 people?*
>
> BAR MANAGER: *At most 100 people.*
>
> PSIU AGENT: *So the fine is US$4,000.*
>
> BAR MANAGER: *I don't think this type of fine is fair. I was misinformed. Besides, the bar on the other corner of this same street has louder music. Why don't you go there? Next to my house there's a church that makes much more noise. Churches can make all the noise they want and the PSIU doesn't do anything.*
>
> PSIU AGENT: *Unfortunately. I'm only enforcing the law.*

The PSIU agent fills in the fine form. The report includes a sketch of the locations where the agent measured the Bar da Esquina's noise and the environmental noise. If the complaints continue and the PSIU agents find the bar continues to operate above the noise limits, they will close the bar (see Figure 4.6).

After the mediation of the city website, the programmer, the schedule coordinator, printers, maps, liaison, inspection agents, vehicles, the sound level meter, a piece of legislation, a technical standard, the Metropolitan Guard officers, the noise from the streets, the noise from the bar, and the bar manager, a fine finally emerges. Once they are done visiting all the scheduled locations, the inspection

FIGURE 4.6. PSIU form for the second offense. After filling in information about the establishment (name, registry number, activity, and address), the form informs the owner or responsible party that "under the terms of [. . .] 1996 Law 11986, I close the aforementioned establishment for continuing to emit noise above the level permitted by law, determining, through this act, the suspension of all commercial activities, until it meets the summons received in the first offense. Disobedience to this term of closure will imply the administrative closure of the establishment with the sealing of all entrances."

agents return to the PSIU, where they deliver the envelope. On the following days, the programmer adds this data into the system. The Fine Sector registers the fine in the case folder. Ms. Freire sees on her computer screen that "An action has been taken."

Bar da Esquina's owner now owes the city US$4,000 for making too much noise. The PSIU will allow the bar to stay open (without playing loud music, of course) as long as it provides, within sixty days, a technical report from a credited acoustic engineer explaining how the bar is going to correct the problem. As mentioned above, if the owner does nothing and continues with the loud music, the PSIU will levy another fine and eventually shut down his bar. In a matter of weeks, the bar owner receives the fine by mail. Unwilling to pay US$4,000 and outraged that only his bar (out of many on that street) is being punished, the owner decides to fight back. His lawyer submits an appeal to the city trying to break the links between the PSIU's report and the Noise Law:

> The appellant recognizes that the inspection of noise emission measured 72 dB at maximum value. He argues that the object of the complaint was the bar located on the other corner on the same block. He questions the validity of the fine, asserting that the inspector did not follow the technical standard NBR 10151 for noise measurement. For instance, the inspecting agent measured the noise *inside* the establishment. Additionally, the noise produced by the global environment was not taken into account.[14]

The Department of Environmental Quality Control, which decides this type of administrative appeal, requests more information from the PSIU regarding the circumstances of the inspection. The case goes back to the PSIU and is handed to the inspection agents. After reading the lawyer's arguments, the agent writes his response: he did *not* measure the sound inside the bar. It may appear that he did so, but when he entered the bar to talk with the manager, he had already made the measurement. He explains that the complaint was made specifically against Bar da Esquina. Also, he *did* measure the environmental noise, as indicated in the fine form. He concludes by suggesting the City of São Paulo *sustains* the fine. The Department of Environmental Quality Control concurs, and the fine is maintained. The bar owner lost the battle in the administrative sphere. He can now either pay the fine or move onto the juridical sphere to contest the city's administrative sentence at the state level—a possibility we will consider in the next chapter.

---

[14] This quote is a synthesis of the most frequent arguments used by bar owners. In Chapter 5 we will analyze this type of litigation more closely.

Note that, at this point, it is hard to know whether Ms. Freire's problem was solved or not. While this extensive assemblage of actors, spread across documents and offices, is deployed to stabilize the fine and solve her problem, Ms. Freire might conclude that the PSIU is either corrupt or inefficient—perhaps both! The temporal gap between the moment a PSIU officer tells Ms. Freire that "an action has been taken" and the night she has a quiet night of sleep, depends entirely on this administrative flow, on the gradual addition of layers, the careful inscription of administrative actions and, eventually, on the effective distinction between legal and illegal sounds. The state's administrative engine is relatively slow and traceable because it needs to avoid falling into legal embroilments, which could damage this bureaucratic machinery and the role of the state itself. At the same time, as Bruno Latour reminds us, this slowness "precisely form[s] the primary material of justice, the material of that which will perhaps one day protect (us or our loved ones) when they are, alas, faced with fighting the coldest of cold monsters, the state" (Latour 2010, 91).

But besides the to-be-expected administrative slowness, it seems clear that the PSIU has been unable to cope with the demand. According to state prosecutor José Eduardo Ismael Lutti, who will reappear in the next chapter,

> The PSIU does not have a structure necessary for the size of the city of São Paulo. There are missing agents, lack of training and qualified agents. And there is another issue: we received a series of complaints that when the PSIU agent arrives at the facility, or a little before, the sound diminishes. And when they go away, the sound increases, which leads to suspicion. I have concrete evidence of some cases. (Personal communication, January 2013)

This statement suggests that Ms. Freire's suspicion might not be completely off the mark. Perhaps there is something else going on with the PSIU. Bribery seems to be an issue at this institution, and the PSIU's relatively obscure modus operandi (which contrasts with the military police's self-proclaimed transparency) certainly does not help. Besides the fine sector break-ins mentioned above, I heard of cases of staffers in the headquarters giving away information about the scheduled inspection to bar owners. In some cases, when the PSIU director found out about the leaks, the person involved was removed from the PSIU (possibly transferred to another agency). I also heard about police officers leaking the inspection schedule to bar owners. "Many police officers work as bar and nightclub security personnel when they are off duty," explained a former PSIU director. "Once I got an invitation to visit the battalion. When I got there, the police officers in charge of street patrolling *told* me the bars they didn't want the PSIU to inspect. I got really mad and left right away." These types of "disruptions" of the administrative flows are easier to occur here because the PSIU requires scheduling and police protection; the inspections are less flexible and more people know about it beforehand. Additionally, sound does not leave any measurable or tangible trace after it ceases. Either the inspecting agent is there to produce the evidence with the SLM and the fine form, or there is very little the state can do to solve the noise problem.

LEGAL-BUREAUCRATIC ARRANGEMENTS

Peace disturbance and decibels have become the two main channels through which sound-politics operate within administrative flows. Of the two, nuisance is the most difficult to stabilize. As "anti-social" behavior, from barbarism to madness and vulgarity, nuisance has been the hold different technologies of power (colonialism, medicine, aesthetics, etc.) seize to quiet "rowdy" groups. The decibel, on the other hand, is the building block of noise pollution, which allows both science and law to define the "average" or "normal" conditions for the hearing body. While noise-as-nuisance is the channel to discipline based on notions of civility and public order, noise-as-decibel is closer to biopolitics, which Foucault defines as the "set of mechanisms through which the basic biological features of the human species became the object of a political strategy" (Foucault 2007, 16).

Although they are endowed with unique disciplinary power, the São Paulo police lack the necessary resources to monitor spaces in search of excessive noise. The COPOM has more than two-hundred surveillance cameras across the city ("electronic eyes that work 24 hours a day, seven days a week, [with] a range of up to 3 km"),[15] but no microphones. The São Paulo police also lack the legal support to punish noise infractions. Complaints continue stopping short of becoming reports, preventing the institution from taking noise into consideration when allocating resources. As we saw earlier, the environment-crime paradigm is partly an effort by the police and residents to bypass the instability of nuisance by fixing noise more firmly within security matters.

Major Tenório's frustration relates in part to the fact that noise prevents the São Paulo police from deploying its statistical apparatus (as Tenório effusively underlined during our interview). The statistics of number-crunching and taxonomy (e.g., the lack of a specific "noise" category at the COPOM because no law requires the police to measure that) is a major challenge for the São Paulo police.

---

[15] "Tecnologia na PM: equipamentos de ponta na luta contra o crime," *Olhar Digital* (April 17, 2011), http://policiamilitar.sp.gov.br/unidades/47bpmm/noticias/tec_pol_sp.html.

It is a challenge because demands for transparency have become a pre-condition for budgetary allocation. In that sense, sound-politics examines not only how the state circulates the notion of noise-as-incivility through its crime categories, but also how it "fails" to reach the status of modernity by not utilizing statistical data.

The PSIU, on the other hand, has a much more limited disciplinary infrastructure, but larger judiciary authority. This authority is not fully within biopolitics because the agency does not conduct statistical analysis of the frequency and distribution of noise in the city to assess the quality of the environment or the well-being of the population for planning purposes, in an attempt to "achieve the right relationship between the population and the state's resources and possibilities (Foucault 2007, 100). Rather, the agency conducts measurements in decibels to give the state legal stability to punish. If the police have often tried to compensate for their lack of judiciary authority by bringing noise closer to crime, we saw that (under Major Rosado) the PSIU can also compensate for the legal obstacles, increasing its disciplinary force by bringing crime closer to noise control. In following the administrative steps necessary for activating the PSIU against the Bar da Esquina, we saw that the rift separating Ms. Freire's urgency in solving her problem and the PSIU's administrative pace puts into question the nature and quality of such judiciary power—citizens know it is there, but they just do not see it.

The PSIU's instability as an institution has hindered any statistical approach to sound-politics. The argument is circular: administration after administration has shifted the agency's modus operandi because it was not structured according to extensive statistical analysis. This allowed the executive branch to change the agency's targets and punishment potential and establish ad hoc negotiation with the legislature (as discussed in Chapter 3). Here again, the slippery condition of sound-politics helps us understand the PSIU's institutional inconstancy. The PSIU is neither a department nor a full-fledged public body, but a "program." The ordinances the agency enforces can punish a wide variety of sound sources. Without a clear target, the PSIU has little institutional grounding; or rather, its grounding becomes more dependent on political arrangements in the Chamber, in the city administration, and in the relation between the two. In the rooms occupied by the PSIU officers, inspecting agents, computers, documents, archives, and SLMs, the "politics" of politicians and their noise circles is particularly tangible.

Public administration specialists explain that the first systematic effort to overcome the clientelism predominant in the country since independence occurred during the Vargas regime (1930–1945). His government made a clear attempt to become bureaucratic in the Weberian sense of the word: hierarchical, meritocratic, standardized, impersonal, rational, and based on a clear separation between public and private spheres. The creation of the Administrative Department of Public Service in 1938, focusing on implementing new administrative methods and increasing state interventionism (as a necessary adjustment to deal with the economic crisis), served as a platform for this project (Costa 2008, 836).

With the return to representative democracy (1945–1964), state leaders started to hire without proper public tendering, which many consider critical for the maintenance of impartiality. In other words, the Weberian ideal was soon "corrupted and dominated by patrimonalist practices widely ingrained [in the country]" (Torres 2004, 147). During the military dictatorship (1964–1985), the bureaucratic and technocratic governance reemerged. However, differently from the Vargas administration approach, in the 1960s and 1970s the state further spread its operations (to expand intervention) and explored private-public partnerships—again, often circumventing public tendering. In a book provocatively titled *Brazil Is Not for Amateurs: Patterns of Governance in the Land of "Jeitinho,"*[16] Belmiro Castor defines these public companies as "devoid of any pressure to efficiency or efficacy because they would be profitable no matter what, simply because the state determined so and, for that reason, imposed monopolistic prices, created privileges, and punished potential competitors with prohibitions and taxes" (Castor 2004, 162).

In the New Republic period (1985–present), a series of privatizing projects generated a range of controversies—"the dismantling of the welfare state," "the neoliberal invasion of multinational corporatism," some claim; "the righteous end of a clogged and wasteful bureaucracy," assert others. The government tried to regain control of its spending by eliminating the foundations and public and mixed-economy companies that had proliferated in the previous two decades. As Frederico Lustosa da Costa explains, while the 1988 Constitution widened mechanisms of political and social inclusion, it "eliminated the flexibility of government-linked companies that, despite their inefficiency and localized abuses [. . .], were the dynamic sector of public "administration" (Costa 2008, 858).

For economist Luis Carlos Bresser Pereira, rather than advancing to a more "progressive" paradigm, the post-dictatorship politicians in office brought the

---

[16] "Jeitinho" (literally "little way") has become a cultural trope to define Brazilian society. Wikipedia explains its meaning as "finding a way to accomplish something by circumventing or bending the rules or social conventions. Most times it is harmless, made for basic ordinary opportunistic advantages, as gatecrashing a party just to get free food and beverages. But sometimes it is used for questionable, serious violations, where an individual can use emotional resources, blackmail, family ties, promises, rewards, or money to obtain (sometimes illegal) favors or to get advantage." https://en.wikipedia.org/wiki/Jeitinho.

administrative machinery back to the Weberian model of the 1930s. For the author, the bureaucratic administration should be invigorated by what the defines as the "managerial" approach. Inspired by the New Public Management developed in the UK in the 1980s, this paradigm focuses on "results" rather than "processes." In the 1990s, under Bresser Pereira's mentoring, the Fernando Henrique Cardoso administration again decentralized the federal government by expanding financially and operationally autonomous agencies (to be monitored by the state through management contracts) and privatizing public institutions. Cardoso's second term in office (1998–2002) focused on reducing public personnel spending while maintaining well-paid career tracks in strategic areas (Gaetani and Heredia 2002).

One of the sources Bresser Pereira draws on to promote his managerial model in Brazil is David Osborne and Ted Gaebler's influential *Reinventing Government*. The book summarizes many of the ideas circulating in the early 1990s among economists and bureaucrats about the future of administrative flows. While the authors (like Bresser Pereira) establish a distinction between their project and the proposal of "running the government like a business" (Osborne and Gaebler 1993, 20), the bottom line is that governments would benefit by incorporating market mechanisms rather than bureaucratic principles. According to Osborne and Gaebler, if the Weberian bureaucratic model has been marked by hierarchical authority, functional specialization, process regulation (emphasis on *how* things should be done), the new entrepreneurial model focuses on "empower[ing] citizens rather than simply serving them" (Osborne and Gaebler 1993, 15). Like Bresser Pereira's managerial approach, the entrepreneurial paradigm embraces private-public partnerships and relies on assessment of performance by stimulating competition.

It is not hard to relate this back-and-forth movement with the institutional controversies we saw inside the military police and the PSIU. The fuzzy gesture of administrative expansion and contraction, of doing more and doing less, centralizing and decentralizing, is ingrained in the institutions that constitute the public administration. If the legal-scientific actors work hard to smooth out the administrative flows by increasing transparency and traceability, they still have to deal with the expansion-contraction spasms that have marked (shall we say plagued?) the executive branch in Brazil. The constant reconfiguration of administrative regulatory measures increases tensions in some areas of the executive branch and generates a level of uncertainty—what is the point of innovating if the incentive may as well disappear in just a few years, and the whole machinery revert to bureaucratic flows (or worse, clientelism)?

BUDGETARY CONCERNS

Between 2012 and 2016, the State of São Paulo invested yearly roughly US$2.4 (5.6% of the total budget),[17] US$2.5 (5.2%),[18] US$2.8 (4.9%),[19] US$3.1 (5.8%),[20] and US$3 (5.7%)[21] billion in public security, respectively. More than half of the public security budget goes to the military police, and investigative and forensic work remains underinvested. Between 2013 and 2016, the state spent at least US$ 21 (R$77) million with stun grenades, tear gas grenades, robotic exoskeletons, and Israeli Wolf Armored Vehicles.[22] When asked why the expenses were much higher than investments in education and culture, the public security secretary explained that "the resources are adequate to maintain the commitments assumed with personnel expenses for an effective force of more than 110,000 police officers, in addition to the maintenance of equipment and materials for all professionals."[23]

At the city level from 2003 (when the Suplicy administration relocated the PSIU) to 2015, the Secretary of Subprefectures has received between US$30 (R$108) and US$110 (R$398) million every year, which represents between 0.6% and 1.9% of the total budget.[24] Besides operating within a budget, institutions such as the PSIU have the potential of bringing money to the city through fines. From 2009 to 2015, it has generated between US$ 5.1 and US$7.7 million in fines for the city every year. In this case, the fine amount is another indicator of efficiency. To increase that number, the city administration might, for instance, stop sending warning letters when there is a complaint or tell the inspection agents not to make informative visits. But taken to the extreme, this approach might backfire and generate political tensions. Citizens are often aware that administrative fines are a source of financial gain for the city. They might argue that the administration is simply interested in profiting from the population with "abusive fines."

---

[17] Fórum Brasileiro de Segurança Publica 2014, 51.
[18] Ibid.
[19] Fórum Brasileiro de Segurança Pública 2016, 61.
[20] Ibid.
[21] Fórum Brasileiro de Segurança Publica 2017, 69-70.
[22] Em 3 anos, gasto com arsenal 'anti-tumulto' em SP chega a R$ 77 milhões," *Folha de São Paulo* (January 25, 2016), http://www1.folha.uol.com.br/cotidiano/2016/01/1733132-em-3-anos-gasto-com-arsenal-anti-tumulto-em-sp-chega-a-r-77-milhoes.shtml
[23] "Governo de SP gasta com materiais de segurança três vezes valor de educação e cultura juntos," *R7* (May 21, 2017), http://noticias.r7.com/sao-paulo/governo-de-sp-gasta-com-materiais-de-seguranca-tres-vezes-valor-de-educacao-e-cultura-juntos-21052017.
[24] 2003: R$108.3 million (R$10.9 billion total budget)—0.9%; 2004: R$120.6 million (R$12.8 billion)—0.9%; 2005: R$110 million (R$12.9 billion)—0.85%; 2006: R$274 million (R$15.1 billion)—1.8%; 2007: R$189 million (R$18.3 billion)—1.03%; 2008: R$253.6 million (R$22.6 billion)—1.1%; 2009: R$149.4 million (R$23.5 billion)—0.6%; 2010: R$334.5 million (R$27 billion)—1.23% 2011: R$530.1 million (R$30.3 billion)—1.7%; 2012: R$659 million (R$34.6 billion)—1.9%; 2013: R$398 million (R$36.7 billion)—1.08%; 2014: R$332.7 million (R$41.6 billion)—0.8%; 2015: R$310 million (R$44.2 billion)—0.7%.

As a device mediating palpable social demands and the economic contingencies of the market (an entity "out there" always hovering over us), the government budget has moved from concerns over controlling spending (the liberal paradigm) to an issue of fiscal policy, redistribution, development, and accountability. Public institutions are increasingly pressured to function not only according to legal parameters that justify their existence but also through the constant analysis of capital gains and losses. Their goal is to "test governmental action, gauge its validity, and to object to activities of the public authorities on the grounds of their abuses, excesses, futility, and wasteful expenditure" (Foucault 2008, 246). The transparency, efficiency, legality, and coordination of an institution's administrative flows has become crucial to justify the allocation and use of public money. This is how one military police captain explained what his institution does: "Today we need to operate both in terms of product and provision of services. My product is crime prevention. The service I provide is attending to and solving the problem. This is how we operate. This is how any company should operate. You need to believe in the product you create, and you need to deliver a good service. And you need customer support and aftersales, so to speak" (personal communication, July 2012).

Osborne and Gaebler argue that the entrepreneurial model turns the city "into money makers rather than budget busters" (Osborne and Gaebler 1993, 18). For the authors, the model is a third route, different from the "liberal call for administrative programs and the conservative call for government to stay out of the marketplace" (Osborne and Gaebler 1993, 284). It is the feedback speed and decentralized nature of markets that governments should incorporate, as this allows the computation of the millions of inputs from citizens more effectively. The problem of the bureaucratic model is that budgets are driven according to constituencies instead of customers, politics instead of policy. This creates a turf mentality and a sense of entitlement among agencies; they rarely self-correct, rarely die, and tend to use commands rather than incentives (Osborne and Gaebler 1993, 287–288). What Osborne and Gaebler suggest, and what our military police captain is alluding to, is a more dynamic approach to administrative flows, where consumer-citizens evaluate the government with regular and updated information.

In Brazil, the government budget at the federal, state, and city level is determined according to three pieces of legislation. The first and more comprehensive one requires the elected administration to send, in its second year in office, a four-year budget plan to the legislative house. This plan determines how much the administration will invest in each public issue (education, health, sports, culture, etc.). It includes a strategic section (a discussion of the current economic context, the goals established by the executive branch, and anticipated funds) and the programs (the problems to be solved and the set of measures to be taken to solve them). The second piece of legislation relates to the planning and the formulation of the budget, a middle point between the four-year budget and the yearly allocation of funds. The third legislation determines each year the funds to be invested in all public institutions (staff, debt, projects, equipment, etc.).[25] Between the 1985 redemocratization and the implementation of the Real Plan in 1994, budgetary decisions were attached to a series of ill-fated overarching economic projects to contain the country's hyperinflation (Castor 2004).

The legislature's amendments to the budget have become a common source of political bargaining. However, these decisions are usually disrupted as the executive sector proposes ad hoc changes based on "emergency" or "relevant" factors. Additionally, the budget legislation *authorizes* rather than *prescribes* the allocation of funds, which gives the executive sector considerable autonomy to make changes as it sees fit. Governors are aware that vetoing the amendments to the budget plan proposed by the Congress risks undermining future legislative projects. From that negotiation, what should be a concise budget proposal with clearly delineated spending goals ends up "pulverizing the scarce resources into millions of projects and activities of political interest to the congresspeople" (Machado 2002, 54; see also Pereira and Mueller 2002).

But the most damaging source of budgetary instability in Brazil relates to two layers of corruption: one being a corruption of the lawmaking process, the other a corruption of the law itself. The first layer relates to clientelism as the strategic partition of public funds to help lawmakers fulfill their personal goals. In the previous chapter we saw that the fate of religious noise in São Paulo was negotiated because mayor Suplicy wanted to create new secretaries and positions. The other layer relates to the series of fissures in the administrative flows that compromise the efficacy of budgetary planning even more drastically. This includes police officers profiting on the side from bar owners to keep the PSIU "in check"; bribed the PSIU officials who leak details about upcoming inspections; municipal officials giving construction companies tax discounts for bribes; overbilling of major construction projects[26]; public and private buildings and infrastructure collapsing because of bribed inspectors[27]; and bribed municipal inspectors failing to fine companies for breaking the Clean City Law. With all these detours, not only is the city prevented

---

[25] Some investments are "locked" into the budget. For instance, the states need to allocate at least 12% of tax revenue for public health and 25% for education; the cities 15% and 25%, respectively.
[26] "Promotoria acusa executivo e fiscais por escândalo do ISS," *Folha de São Paulo* (April 5, 2017), B3.
[27] "Prefeitura apura se propina liberou prédio que desabou," *Folha de São Paulo* (August 30, 2013), C4.

from profiting from fines (corrupted officials operate in the quantifiable difference between fine and bribe), but the private sector (and society more broadly) becomes regulated by an alternative channel. The public service is financially and morally emptied out (even when it is fully present in the planned budget), and citizens see and engage with the law-enforcement actors as a mere pretense of government. Whether the modus operandi follows the bureaucratic (where top technocrats control vast bribery networks) or the managerial model (where competing companies negotiate behind the scene the prices for public contracts), clientelism remains largely dominant in Brazil. For many citizens, budgetary decisions continue to follow the perverse plutocracy of rotten politicians.

PARTY AFFILIATION

The third source of administrative instability brings us closer to the political circles. Castor suggests that even when institutions have a good number of public officers operating under the bureaucratic principle of meritocracy and neutrality, they are headed by groups operating under a patrimonial structure that runs across party lines (Castor 2004, 140). According to Gaetani and Heredia, by 1988, there existed at least four types of public personnel regimes within the state apparatus: first, institutions known as pockets of excellence where public sector employees were recruited through private contracts on meritocratic grounds; second, government employees recruited through private contracts on a patronage basis; third, civil servants recruited on a merit basis like the diplomats and tax inspectors; fourth, civil servants who acquired tenure and pension rights through political or administrative means (Gaetani and Heredia 2002, 8).

Visit any public institution in Brazil, and chances are the group of people you see working represent a mix of these different career regimes. As you move higher in the hierarchy of an institution, you will see more commissioned positions. "At every election, the scene of commissioned officers seeking support of influential politicians to keep their jobs repeats," Castor explains, "with negotiations whose consequences are easy to imagine" (Castor 2004, 141). This is more often the case with civil institutions such as the PSIU than with military ones such as the police. According to state prosecutor Lutti, "There's a lot of political interference in this institutional body. And then it becomes difficult for the employee who wants to work, with so much political interference" (personal communication, January 2013). Since 2012, when I first visited the PSIU, commissioned personnel have come and gone depending on the party heading the administration.

In my last visit in June 2017, nine of the roughly twenty people working at the PSIU headquarters were gone. This affected the tendered officers, as they had to fill in other positions. "The person who was working at fine sector is gone. She was supporting the opposing candidate in the elections," explained one of the tendered PSIU officers matter-of-factly. Every four years an administrative ebb and flow takes place across public institutions, where the "troublemakers" are dismissed or moved to harmless positions and the "loyalists" are brought in to create a reliable communication channel between the mayor's office, his or her secretaries, and the institutions.

Scott Mainwaring notes that Brazil has been marked by "considerable instability in patterns of party competition, weak party roots in society, comparatively low legitimacy of parties, and weak party organizations" (Mainwaring 1999, 3). Politicians have limited loyalty and discipline, changing parties (or creating new ones) with such regularity that people often lose track of which party a candidate is affiliated with. With little ideological attachment, politicians change parties whenever they envision better chances to get elected or be invited to a position in the administration.[28] There is a tacit understanding that clientelism is the real currency. The so-called social cleavage approach, which maintains that "social identities such as class, religion, ethnicity, and region provide the basis for common interest and thereby create enduring partisan sympathies" (Mainwaring 1999, 21) has not materialized in Brazil. Rather, groups united by social identity (religious affiliation in particular) vote together but operate across partisan lines to increase lobbying potential and chances to hold strategic positions.

Party oscillation has been particularly discernible in São Paulo. Since redemocratization in 1985, no mayor has managed to get reelected. In the last fifteen years, party affiliation in the city has been divided between the center-left PT and the center-right PSDB, each supported by dozens of satellite parties. As mentioned in the previous chapter, the center-leftist PT emerged in the late 1970s as a grassroots movement among union leaders, activists, and intellectuals. With administrations in 1989–1992 (Erundina), 2001–2004 (Suplicy), and 2013–2016 (Haddad), the party has grown from a platform of participatory and decentralized citizenship. This includes, for instance, attempts to institute a more participatory budget,[29] strengthening the subprefecture system, and creating cultural opportunities in the suburbs (cultural citizenship has been a mark of the party

---

[28] "In the Brazilian legislature of 1991–1994, the 503 deputies changed parties 260 times" (Mainwaring 1999, 37).

[29] For a discussion of the public hearings related to the budget voting at the São Paulo Municipal Chamber, see Brelàz and Alves (2013).

since the Erundina administration). It was during the PT administration that the city passed its two last Master Plan laws (in 2002 and 2014) and zoning laws (in 2004 and 2016). This put the PT at the center of debates about the right to the city, which has pitched disenfranchised communities of insurgent citizenship (the homeless and the suburban poor) against the elites of differentiated citizenship (real estate developers in particular). However, this can prove to be a risky political approach. As the prelude to Chapter 3 suggests, when the PT is not able to satisfy the demands of activist groups due to budget concerns, this can empty the party's support system and put it in a fragile position.[30]

Following their party ideology, PT mayors have proposed to decentralize public institutions under the premise of bringing it closer to the average citizen. They have also nominated technocrats to direct the agency, who were willing to follow the mayor's political agenda away from zero tolerance.[31] The interviews with the former PSIU officials suggest the most common criticisms against the PT's decentralizing impulse: first, the lack of a standardized treatment to law enforcement, as each subprefect might have an understanding of the law; second, lack of organization (with documents circulating across a wider spatial network, it becomes difficult to know who is doing what); third, the increase of paperwork; fourth, it increases the chance of corruption and local ad hoc negotiations.

We already considered the impact of the subprefecture system on the PSIU in the early 2000s. This shift resonates with the sources of administrative instability mentioned above as well. The change was intended to increase territorial autonomy and answer to the citizens' demands more promptly. One major bureaucratic issue was the division of responsibilities among the centralized municipal secretaries and the decentralized subprefectures. The Suplicy administration structured the subprefecture into seven departments.[32] As most of these departments encompassed more than one municipal secretary (there were twenty-one secretaries), this change created an administrative impasse between the municipal secretaries (who wanted to retain as much delegating power as possible) and the subprefects.

The lack of a standardized distribution of tasks raised budgetary concerns. Three departments (Education, Maintenance and Infrastructure, and Health) represented 20% of the city budget, but 96.31% of the subprefecture resources on

---

[30] A good example of this situation was the massive protests against the bus fare increase in 2013, during Haddad's first year in office (see Romero 2013).
[31] This has caused many political tensions, as the military and civil police work for the governor.
[32] Social Action and Development, Management and Finances, Urban Planning and Development, Urban Maintenance and Infrastructure, Projects and New Works (*Projetos e Obras Novas*), Education, and Health.

average (Grin 2015, 135). Subprefects complained that the investments coming from the secretaries (i.e., what had not been decentralized) did not coincide with the local demands. Fearing the autonomy of the subprefectures and loss of political control, secretaries started to bypass the subprefect and negotiate the budget directly with the heads of the subprefecture departments (Grin 2015, 135).

The impact of the subprefecture system on party alliances was another major point of controversy. In the era of Regional Administrations (1965–2001), city councilors were the main administrative channel between community and state. They used the Regional Administration as a means to establish and maintain their political/electoral platform. The new subprefecture model created tensions with the Municipal Chamber because it gave the executive sector autonomy to distribute positions across the subprefectures. The frictions between subprefecture and secretary created disagreements within the PT as well, as the Suplicy administration used the creation of new positions in the subprefectures as a bargaining mechanism with the Chamber. As the 2003 Secretary of the Department of Subprefectures would later state, "We can say that the Suplicy administration built a parliamentary majority, and the priority terrain of the agreements was the subprefectures" (quoted in Grin 2015, 138).

The other popular party is the PSDB, a party that emerged in São Paulo in 1988 from dissidents of the centralist, heterogeneous, and hegemonic PMDB (Party of the Brazilian Democratic Movement). Thus, unlike the PT (and many European social democratic parties), the PSDB was created exclusively by politicians (most of them based in São Paulo). As Celso Roma argues, although the origin of the party is often narrated as a matter of ideological differences with the right-wing parties, a more plausible reason is the pragmatic decision of establishing a new party to create political opportunities—particularly in the upcoming 1989 elections (Roma 2002).

The PSDB is more closely connected to business interests, public security, property rights, and individual liberty. It assumed a center-rightist position, defending economical changes and a competitive market as solutions to both populism (from the right) and statism (from the left). These changes include economic deregulation, privatization of state companies, relaxation of labor laws, and separation of state and labor unions. With administrations in 2005–2006 (Serra), 2006–2012 (Kassab),[33] 2017–2018 (Doria), and 2018– (Covas), the PSDB has embraced zero tolerance and relied more often on centralized disciplinary mechanisms to punish

---

[33] Kassab is not a member of the PSDB. I include him in this group loosely—he was Serra's vice mayor and won the 2008 elections with support from the PSDB.

illicit behavior. It is not surprising then that PSDB mayors have appointed people with extensive police background to direct the PSIU.

If the new subprefecture system was a way for us to consider the relations between administrative flows and the three sources of instability (legal bureaucratic arrangements, budgetary concerns, and party affiliations) in the center-left, the Delegated Operation provides an interesting case study to see a similar dynamic in the center-right. As mentioned above, the main purpose of this initiative was to eliminate informal commerce in the city and increase public security in the streets. Soon the presence of off-duty police officers participating in the Delegated Operation expanded to other subprefectures. Employing the politics of numbers, shortly after implementing the operation, the city administration stated that robbery had diminished 20% and vehicle theft 29%.[34] Besides nominating retired colonels as subprefects, the Kassab administration further militarized the city by hiring police officers to work as municipal officials. The governor of São Paulo, a member of the PSDB, not only praised the success of the Delegated Operation to reduce crime but announced its expansion to other cities. The mayor also raised economic concerns to justify the presence of the police in the streets: "Illegal merchandise takes the job of a family father. In addition to committing this crime, these people fail to pay taxes, and unfortunately, they end up contributing to the municipal administration's failure to build schools and health centers."[35] To get enough police officers in the operation and refute worries that the city would not be able to fulfill its budgetary obligations, Kassab agreed to transfer the funds to the military police every two weeks.[36]

As the link in the administrative agreement between city and state, the Department of Subprefectures could technically instruct the police-inspectors to enforce other tasks under its jurisdiction. In the PT administration, the city extended the Delegated Operation to enforce the 1:00 a.m. Law with the PSIU. The chief of the military police, Colonel Camilo (who later joined Kassab's political party), declared that the PT administration had "destroyed the Delegated Operation."[37] For him, besides reducing the number of hires and detracting from

---

[34] "SP amplia fiscalização contra o comércio ilegal nas zonas sul, norte e oeste," *Folha de São Paulo* (May 26, 2010), http://www1.folha.uol.com.br/cotidiano/741075-sp-amplia-fiscalizacao-contra-o-comercio-ilegal-nas-zonas-sul-norte-e-oeste.shtml.

[35] "Dois milhões de produtos pirtas são destruídos em SP," *Folha de São Paulo* (July 20, 2010), http://www1.folha.uol.com.br/fsp/mercado/me2007201016.htm.

[36] "Prefeitura de SP atrasa pagamento de 'bico oficial' de PMs," *Folha de São Paulo* (February 15, 2017), http://www1.folha.uol.com.br/cotidiano/2013/02/1231135-prefeitura-de-sp-atrasa-pagamento-de-bico-oficial-de-pms.shtml.

[37] "A Prefeitura destruiu a Operação Delegada," afirma Camilo, *PSD São Paulo* (April 25, 2016), http://psd-sp.org.br/saopaulo/prefeitura-destruiu-operacao-delegada-afirma-camilo/.

the original purpose (controlling illicit downtown commerce), the PT had failed to transfer the payment to the military police on time. Police officers refused to work in the peripheries enforcing the 1:00 a.m. Law. "No one wants to work far from home, especially at night," explained one police officer.[38] In 2014, a police officer was arrested for killing a hawker during a Delegated Operation. The military police informed that the hawker was shot after he tried to take the officer's gun. They also stated that the officer was wounded during the altercation. After footage of the event circulated in the press, showing the hawker trying to take the pepper spray from the officer, the institution stated it was not so sure about the facts.[39] In 2017, the city councilors hired thirty military police officers through the Delegated Operation to protect the Chamber.[40]

The coordination and extension of the administrative flows within the military police and the PSIU considered in this chapter are affected by these instabilities, which in turn are constantly interfering with each other. In considering these flows in the first part of the chapter, we saw that, although citizens believe in the government's ability to eliminate unwanted sounds (as the high number of calls received by both institutions indicate), many complaints end up in administrative limbo. Actors operating within both institutions understand the disjunctive nature of noise control in the city and attempt to either circumvent these discrete boundaries or reassociate the issue of noise with other (more stabilized) problems such as crime prevention.

To be sure, there are other sources of instability in the executive sector we could consider, such as corruption scandals and media pressure. But these would be either amplifications or extensions of the three main issues that most often demand constant deliberation and that end up defining an administration. In examining the executive sector as a field organized through administrative flows, we have reached a central node in the book. This node is particularly relevant because it can use legal-bureaucratic principles, budgetary concerns, and party affiliations, to drastically shift how a public institution deals with noise. Rather than considering institutions such as the military police and the PSIU as homogeneous institutions where officials are expected to follow standardized protocols, we localized the very notion of a "standardized protocol" within public administration debates. We also

---

[38] "Policiais recusam 'bico oficial' em bairros da periferia," *Folha de São Paulo* (May 10, 2013), http://www1.folha.uol.com.br/fsp/cotidiano/108134-policiais-recusam-bico-oficial-em-bairros-da-periferia.shtml.
[39] "PM é preso após morte de camelô durante operação na zona oeste de SP," *Folha de São Paulo* (September 18, 2014), http://www1.folha.uol.com.br/cotidiano/2014/09/1518328-pm-e-preso-apos-morte-de-camelo-durante-operacao-na-zona-oeste-de-sp.shtml.
[40] "Câmara Municipal de SP pagará PMs por 'bico oficial' no entorno do prédio," *Folha de São Paulo* (June 5, 2017), http://www1.folha.uol.com.br/cotidiano/2017/06/1890255-camara-municipal-de-sp-pagara-pms-por-bico-oficial-no-entorno-do-predio.shtml.

saw that both institutions have had a certain level of fluidity and inner tension, with key actors working to either increase or decrease, thicken or narrow, the administrative flows that mediate authoritative documents and a range of events. Administrative flows *modulate* according to different modes of existence—chains of reference, technological folds, political circles, and legal channels. This puts enormous pressure on, but gives considerable resources to, the city administration. As Mainwaring notes, "Winning executive power sets the whole policy agenda; social movements and interest groups affect specific items on the agenda" (Mainwaring 1999, 13). But setting the policy agenda and having the opportunity to rearrange legal-bureaucratic ties, budgets, and party affiliations do not give the administration unlimited authority. In the next chapter, we will consider some of the actors that can monitor and challenge the executive sector's back-and-forth movement from regulatory documents to "feral belongings."

5

LEGAL CHANNELS

FROM THE PSIU headquarters on Libero Badaró Street, if you walk one block toward City Hall and take a left on Quitanda Street and a right on November 15 Street, you will eventually catch sight of a white colonial mission, the place where Jesuit priests founded the city in 1554. Continue walking on November 15 Street and you will end at the *Sé Square*, the city's historic central plaza. On the opposite end of the plaza, you will see the imposing Sé Cathedral, a 300-foot church built in the neo-Gothic style between the 1910s and 1950s. Next to the cathedral, a few feet to your left, you will see another majestic building, the *Palácio da Justiça*, or the "Palace of Justice." The building, inspired by the Roman Palace of Justice, was planned in the 1910s by Francisco Ramos de Azevedo, the famous architect who revamped downtown São Paulo according to the architectural eclecticism then popular in Europe and North America.

The *Palácio da Justiça* is the command center of the São Paulo state judiciary branch. It is here that roughly 360 appellate judges analyze, debate, and decide thousands of cases every day (about one-third of all ongoing legal cases in Brazil).[1] The São Paulo Court of Justice "is considered the largest court in the world in terms

---

[1] There were 18 million legal actions in 2009 alone. "Judiciário de SP tem mais de 18 milhões de ações," *Consultor Jurídico* (March 5, 2009), http://www.conjur.com.br/2009-mar-05/judiciario-paulista-18-milhoes-processos-julgar.

FIGURE 5.1. Aerial view of São Paulo's administrative center, also known as "historic downtown," showing some of the administrative buildings visited during fieldwork.

Credit: Google Earth.

of its volume of cases."² The Court is led by Supreme Judiciary Council (*Conselho Superior da Magistratura*), which includes the presidents of the Criminal, Public, and Private Law sections. The Private Law section, the largest, has 190 appellate judges organized in almost forty chambers—some of which have specialties in areas such as brands and patents, bank contracts, and property law. The Public Law section has ninety judges working in eighteen chambers, three of which deliberate on city taxation and two over workplace accidents. The Criminal Law section has seventy-nine judges divided into sixteen chambers. Since 2005, the Court has two chambers reserved to environmental litigation, which are made up of judges from the Criminal, Public, and Private Law spheres. It is in this building that most of this chapter unfolds. (See Figure 5.1.)

After following newspaper stories, acousticians, politicians, police officers, and noise inspectors, we must now meet the legal experts. To be sure, this is not the first time that we come across the intersection of sound and law. Remember, for instance, the heated debates concerning the legal ramifications of the ABNT technical standards (Chapter 2), the controversy about the constitutionality of noise legislation in addressing religious services (Chapter 3), or the legal conditions determining how, when, and where the police and PSIU could operate (Chapter 4). Each of these issues pointed to a set of actions, sounds, documents, and people that were scrutinized according to the relations between legal parameters *and* something else—scientific objectivity, political circles, and administrative flows. In those debates, law was only approached tangentially, hardly allowing us to grasp its specific validation mode. Such a tangential approach to law is not unusual. As Latour notes, "the legal institution is so porous that its decisions look like so many weathervanes, turning with every breeze" (Latour 2013, 362). This fluidity seems to contrast with the set of values it is supposed to protect so rigidly. The contrast, however, is what makes this mode of existence so unique. It is precisely this flexibility to incorporate virtually any controversy that allows any problem to slowly (through dossiers, clerks, lawyers, judges, and experts) conform more and more to the legal mode of existence.

It is time to delve into this network and learn how legal specialists shape sound-politics by converting environmental noise into a set of encoded principles. This conversion is necessarily delicate because it draws on and mobilizes all the other networks discussed so far—technical standards, laws, and administrative flows. Successful legal action requires not only the just interpretation of laws but also the constant assistance of scientific measurability and the executive branch's authority to inform and enforce judicial decisions through unbiased agents. It is by

---

² "Quem Somos," *Tribunal de Justiça, Estado de São Paulo* (n.d.), http://www.tjsp.jus.br/QuemSomos.

attaching these scientific networks and administrative flows that a legal case gains *materiality*—that is, the evidence necessary to turn an allegation into a fact.

This chapter examines five legal channels related to environmental noise litigation. The first channel, unlike the other four, is within public law. Here, I will focus on the appeals against the PSIU's administrative fine. In following the Bar da Esquina case in Chapter 4, we saw that after losing the administrative appeal, the convicted bar owner could still appeal to the judiciary to overturn the city's fine. I consider that and other cases as they move from the municipal to the state level. Aside from the first channel, we are going to move from private to diffuse rights, and from light to serious infractions in the civil and criminal spheres. The second channel invokes the Civil Code and relates to cases in private law known in Brazil as "neighborhood law" (*direito de vizinhança*)—the collection of rules that regulate property use to mediate conflicts between occupants of neighboring buildings. The other channels involve the Public Prosecutor's Office (*Ministério Público*, or MP). The third channel is the *Public Civil Action*, in which the MP tries to stop excessive noise in the name of collective interests. The fourth and fifth channels bring us to criminal law: I examine litigation surrounding misdemeanor law and the Environmental Crime Law.

We are clearly dealing with an issue that crosses a range of legal fields. The heterogeneity and evasiveness of noise, identified in the hybrid forums, political circles, and administrative flows, is visible here as well. Depending on the sound source, the legislation deployed, and the inclination of the complainant, lawyer, and judge, noise controversies can land on different desks in the Court of Justice—sometimes activating different legal fields concurrently. The work ahead is challenging, to say the least (especially when the ethnographer is not a legal specialist!). To simplify things, I limit myself here to appeals at the São Paulo Court of Justice. I do this for a couple of reasons. First, there is the issue of ethnographic practicality, as all cases were decided in one location. Second, and most obviously, it reduces the number of cases to be analyzed. This makes it easier for us to spot recurring issues and thereby delineate the Court of Justice's jurisprudence related to noise. Finally, focusing on appeals allows me to include legal actions taken against the PSIU and consider the city as a litigious party rather than an arbiter.

The chapter is organized into two main sections. The first part focuses on three crucial translators in this network: the lawyer, the state prosecutor, and the appellate judge. These agents will guide us through the series of procedural steps necessary for moving along the legal channels. This will give us a sense of what types of challenges these three professionals usually encounter when adding materiality to a case, analyzing evidence, interpreting legislation, and linking facts to principles.

Once we have understood how these actors build and make decisions on noise litigation (step-by-step, document-by-document), we can then approach the five legal channels more fully. The second section focuses on jurisprudence. I discuss how cases enter the Palace of Justice as legal appeals and leave the building as either the confirmation or rebuttal of the administrative fine and initial judgment sentences. The section examines the weight of legal documents in the Court of Appeals and the most frequent strategies the parties use to articulate the law to their own advantage.

THE LEGAL ORDEAL

The Lawyer

Besides being a well-established lawyer in Private Law (specifically landlord-tenant litigation), Waldir de Arruda Miranda Carneiro is a legal scholar, a group known in Brazil as *doutrinadores*. The doutrinadores have the important task of helping lawyers and judges interpret the law by tying together codes, the Constitution, legislation, judicial decisions, and scientific studies. Carneiro became interested in noise around 1999, after buying an apartment with poor soundproofing, which led him to sue the construction company. "Here in Brazil, we don't follow the American technical specifications for construction," he said. "If you make a hole in your drywall, you are going to see your neighbor on the other side!" (personal communication, March 2012). In 2001, when he first published *Perturbações Sonoras nas Edificações Urbanas* ("Sonic Disturbances in Urban Buildings"), a book that quickly became a reference in community noise litigation, Carneiro could only find a few dozen legal cases related to neighborhood noise in Brazilian courts. By the fourth edition of his book (Carneiro 2014), he had found more than 400.

Those in the Anglosphere familiar with common law, which is strongly shaped by the *stare decisis* (the premise that precedent decisions having binding authority over subsequent ones), might be puzzled by the role of doutrinadores such as Carneiro in civil law tradition found in Brazil and most of Latin America. This system evolved from ancient Roman civil law, Roman Catholic Church canon law, and commercial law, and was later extended and expanded under the influence of the American and French Revolutions, which inserted notions such as the separation of governmental powers, public law, and constitutional law. In civil law, the doutrinadores not only interpret the law, but help judges apply the law by instituting 1) the means through which a case moves through the process, "which enables the transportation of obligation from one end of the procedure to the other, and from a text to the case at hand" (Latour 2010, 194); 2) the coherence of the law itself, as it shifts from an abstract to a comprehensible and palpable state; and 3) the limits of the law, which is important for preventing both complete paralysis (everyone suing everyone) and extremely restrictive access to the legal system.

Besides the presence of the doutrinadores, there are a few other differences between civil law and common law systems that are worth mentioning. First, the codification practice is a central element in civil law. Codes have precedence over previous decisions. To approach the world of civil law, we thus need to understand the constant flux involving legal coding and decoding, where "the teacher-scholar is the real protagonist" (Merryman and Pérez-Perdomo 2007, 56). Codes usually draw on constitutions and help specialists in court cases by providing more detailed regulatory parameters. The five most important codes in civil law are the Civil Code, the Civil Procedure Code, the Criminal or Penal Code, the Criminal Procedure Code, and the Commercial Code.[3]

Brazil's first Constitution (1824) anticipated the creation of criminal, commercial, and civil codes. Whereas lawmakers managed to approve the Criminal (1830) and Commercial (1850) Codes relatively quickly, it would take them almost a century to pass the first Civil Code.[4] The 1916 code had "only" 1,807 short, fairly abstract, articles. In the 1980s, the revision of the civil code stalled because of the political turmoil that led to the end of the military dictatorship. In 2001, Congress resumed work on the code, and in 2002, the president signed into law the new Brazilian civil code.[5] The most recent civil procedure code became law in 2015,[6] replacing the 1973 version. The current Criminal Code[7] and Criminal Procedure Code,[8] together with the Criminal Contravention Law[9] date back to the Vargas era (1930–1945). Because the codes were created at different moments by lawmakers with different viewpoints, legal specialists navigate an ocean of codified law that has its own inner tensions and conflicts. Adding to this instability, Brazil has had

---

[3] There two important codification legacies. The Napoleonic code of 1804 pushed for easily accessible, understandable, and universal regulations in an attempt to bypass lawyers. In contrast to the French code, the 1896 German civil code was conceived for lawyers (with the use of legal jargon) and took into consideration local norms. Both the German and the French codes supported a strict separation of powers and attempted to be comprehensive by covering a wide range of topics.

[4] Throughout the nineteenth century, several jurists were hired to accomplish this task. In the 1900s, special commissions in the House of Representatives and Senate drew on these previous attempts to pass the first code in 1916. Although the French and the German codes had influences on the Brazilian civil code, its creators followed the Portuguese tradition by drawing heavily on Roman law.

[5] Federal Law 10406/2002

[6] Federal Law 13105/2015.

[7] Federal Decree-Law 2848/1940.

[8] Federal Decree-Law 3689/1941.

[9] Federal Decree-Law 3688/1941.

more constitutions[10] (supposedly more stable due to their importance over all other laws) than civil and criminal codes combined.

Another important difference between civil law and common law relates to the role of the judge. The French separation of powers and codification practice were post-revolutionary attempts to neutralize the abuses of aristocratic judges. This created a "separate system of administrative courts, inhibited the adoption of judicial review of legislation, and limited the judge to a relatively minor role in the legal process" (Merryman and Pérez-Perdomo 2007, 19). Civil law judges are public servants without the same influence as their common-law counterparts on how to interpret and apply the law. Unlike in common law, civil law lawyers and judges follow specific career tracks early on, and there is little professional migration between the two professions.

A final difference worth mentioning relates to legal proceedings. Rather than a single event (the trial in common law), proceedings in civil law are usually fragmented. They include a series of highly formalized written exchanges between the judge and parties. For example, to question a witness, civil law lawyers must submit a written statement to the judge and the opposing lawyer. As Merryman and Pérez-Perdomo observe, "The element of surprise is reduced to a minimum, since each appearance is relatively brief and involves a fairly small part of the total case" (Merryman and Pérez-Perdomo 2007, 114). Because of this diluted format, it is not uncommon for a legal action in civil law to change shape *throughout* the proceeding. The right of appeal in civil law includes the right to introduce new evidence, and appellate judges have the autonomy to consider factual (content) as well as legal (formal) issues.

One can see civil law in action by following Carneiro as he moves from a noise complaint to legal action. To follow this process, I draw on interviews with Carneiro as well as on court documents. He starts the litigation by filing an *initial petition* in a trial court. The document explains that the plaintiffs are neighbors requesting that a restaurant located across the street stop making noise above the limits determined by the technical standard NBR 10151.[11] They ask for compensation for moral damages.[12] The petition explains that the restaurant hosts events with loud music from 7:00 p.m. to 3:00 a.m., five days a week. To avoid having to wait for

---

[10] There are seven constitutions in total: 1824 (post-independence), 1891 (the New Republic), 1934 (Vargas), 1937 (Vargas), 1946 (post-Vargas), 1967 (military dictatorship), and 1988 (return to democracy).

[11] The NBR 10151 nighttime decibel limits in predominantly residential areas are 50 dB (outdoors), 40 dB (indoors, with open windows), and 35 dB (indoors, with closed windows).

[12] In private law, whether the defendant owns commercial or residential property is not relevant for this type of legal action. The plaintiff's argumentation would not differ significantly if the defendant were another neighbor playing loud music in his apartment.

the final sentence to have some peace, the plaintiffs include an injunction request in their petition, asking the judge to order the restaurant to cease making noise immediately.

The judiciary receives the petition and sends a legal officer to deliver the *notificação* to the defendant, giving the restaurant owner fifteen business days to submit a response. In his response, the defendant explains that the restaurant is not at fault, as it is located on an extremely busy avenue. The plaintiff then has another fifteen business days to counter the plaintiff's response. With those documents in hand, the judge then prepares a *conclusive opening order*. In this document, she considers preliminary issues (e.g., whether the parties have a legitimate interest in the case and whether the case falls under her jurisdiction), confirms whether the accusation has litigious potential, and identifies the controversial points to be clarified. Based on the requests made by each party, the conclusive opening order specifies the types of evidence (documents, testimonies, or expert reports) necessary to clarify the controversial points. In most private law cases, the most important evidence is the noise measurement conducted by an expert.

Anticipating these steps, Carneiro includes in his initial petition an expert report that clearly states that the restaurant is exceeding the limits established by NBR 10151. Knowing that this evidence could be considered biased, Carneiro hires a notary public to accompany the measurements. The notary public issues a document, imbued with legal neutrality, that describes the measurement and gives the report credibility. Opening a case with an expert report already on hand is not a cheap option,[13] but it gives the plaintiff enough leeway to persuade the judge to consent to the injunction request. The judge grants the plaintiffs' injunction request and orders the restaurant to respect the NBR 10151 limits or face a US$3,000 fine for each day in which it exceeds the noise limits. A few weeks later, the plaintiffs submit evidence showing that the restaurant continues to play loud music, the judge raises the fine to US$14,000/day. The judge nominates an expert to conduct the noise measurements. She asks each party to provide the expert with a list of questions to be addressed in the report. "These questions are crucial because they conduct the production of proof," explains Carneiro (personal communication, June 2017).

During a trial, the judge issues three main types of documents: rulings, provisional sentences, and the final sentence. *Rulings* are minor administrative decisions that move the case forward (e.g., requesting a report from the judicial expert). The *provisional sentence* is a nonconclusive decision taken during the proceedings. These decisions include, for instance, the injunction determining the

---

[13] Hiring the notary public and the expert costs a few thousand Brazilian reais.

immediate cessation of noise above the legal limits. Injunctions are decided before the final verdict because they address actions that might cause irreparable or barely-reparable harm to a party (such as exposure to excessive noise). The *final sentence* brings the litigation to an end (at least at a given court). The final sentence includes three sections: a summary of the events, the reasoning behind the decision, and the verdict.

In our example, after analyzing the evidence provided, the judge finds the restaurant owner guilty. Confirming her provisional decision, she orders the restaurant to cease making noise above the NBR 10151 decibel limits. The plaintiffs requested compensation for moral damages, arguing that the noise caused psychological and moral injuries. This is a convoluted issue because there are no fixed parameters to quantify psychological injuries. Carneiro explains that, in comparison to the United States, Brazilian judges are usually "restrained" when it comes to defining a value for moral damages. The judge orders the restaurant to pay US$1,000 to each plaintiff for moral damages.

The State Prosecutor

With Carneiro, we learned how to move the process in the Private Law channel. But when noise affects an undefined number of people, then it is the state prosecutor (*promotor de justiça*), as a representative of the collective, who enters the scene. The 1988 Constitution gave the Public Prosecutor's Office (*Ministério Público*, or MP) "functional and institutional autonomy" (Art. 127) to defend "social interests" at the state and federal levels by taking part in criminal and civil litigation. Since then, the MP has grown and become multifaceted. Today, it works on a wide range of issues, with sectors focused on consumer rights, urbanism and environment, public property, children and adolescents, the elderly, elections, and public health. As an independent governmental body, the MP can investigate and take legal action against all three governmental branches.

It is part of state prosecutors' jobs to take public institutions and officials to court for not doing their job. State prosecutor José Eduardo Ismael Lutti, who works in the Urbanism and Environment sector of the São Paulo MP, is critical of the PSIU. For Lutti, "If the PSIU functioned properly—if the city administration functioned properly—many things could be avoided. There is a big problem with law enforcement. Residents call the PSIU, the police, and finally, as last resort, end up making a complaint to us" (personal communication, May 2013).

One of the MP's instruments is the Public Civil Action (Law 7347/1985), which focuses on cases involving diffuse or collective interests—assets obtained through litigation are sent to the state's Diffuse Interests Fund. Anyone can request that the

Public Prosecutor's Office initiate a public civil action. If the MP believes that the case has enough materiality (i.e., that legal form and factual content converge), it will initiate an investigation, requiring expert reports and documents from public agencies. In noise-related litigation, the strategies for adding materiality to a case are not different from those we have seen with Carneiro. As the MP does not represent individual interests, the state prosecutor needs to make sure that there is indeed a group of people affected by the noise. One way to indicate the diffuseness of the issue is to ask complainants to collect signatures from several neighbors.

Once Lutti is confident that there is enough public interest in the case, he initiates a legal action against the person, business, or governmental entity making the noise. "We receive every type of complaint," Lutti explains, "from nightclubs and bars to A/C systems in large properties, to the subway line construction." As we know, to speed up the case and give weight to the injunction, it is a good idea to include an expert report in the initial petition. Lutti explains that he often goes to the CETESB (introduced in Chapter 1) to get an expert report because the PSIU can often be "unresponsive."

Let us consider another case of community noise. How does the state prosecutor deal with a bar that has been exposing neighbors to loud music? Based on the request from residents of two buildings next to the bar, Lutti opens an environmental Public Civil Action against the venue. The prosecutor's initial petition includes documentation of the PSIU inspections and police reports issued against the bar. He also includes the invoice of soundproofed windows that the residents installed, dozens of signatures by neighbors demanding governmental action, and oral testimonies. The prosecutor is careful to explain that the harm the bar inflicts in the area affects not only the identified residents but an *undefined* and *undetermined* number of people. In this way, he clearly repels potential attempts from the defendants to claim this is a private law issue, which is outside the MP's jurisdiction.

The MP requests a provisional sentence ordering the PSIU to close and wall up the bar and take away its operating license immediately, and fine the bar US$30,000 for every day it violates the judicial order. Lutti includes as codefendants not only the bar owners but the PSIU and the military police for failing to enforce the law. He requests that the PSIU or another municipal agency create and organize, in no more than 90 days, a 24/7 noise inspection service available via specific phone line or face a US$30,000 per day fine. From the state, it requests noticeable policing in the region to prevent noise pollution, or the payment of US$14,000 for each time it fails to respond to the community's requests. On top of that, it requests compensation for diffuse moral damages: US$470,000 from the bar for five years of noise (taking as a base the PSIU's fine value per infraction), US$190,000 from the city, and US$94,000 from the state. All defendants should also pay compensation for the soundproofed windows and for any physical and psychological harm caused by the bar's noise.

The MP can move in the criminal law sphere as well. Once the person goes to the police station to file a police report, the civil police conduct the investigation and send the police inquiry report to the judge. As misdemeanor and environmental crimes involve unconditional and public criminal actions,[14] they require that the MP acts as the plaintiff representing the state. In such instances, the judge sends the case to the state prosecutor. After analyzing the case, the state prosecutor can initiate the legal action here as well, or ask the judge to shelve it for lack of materiality.[15] Based on a preliminary formal analysis and the evidence, the judge can decide whether to accede to the state prosecutor's request. If the judge thinks the case should move forward and the MP requests for it to be shelved, the judge will send it to the state appeals prosecutor. If the state appeals prosecutor thinks it should be shelved, then the judge must comply. If the state appeals prosecutor agrees with the judge, then the case will go to another state prosecutor.

This gives us a sense of how the MP intermediates between the executive and judiciary branches, and how it uses both to defend the public interest in the civil and criminal spheres. It is not hard to see why the MP can make these branches uneasy. As (usually) the last legal resort, solicited at a point where disgruntled citizens have decided to combine forces, the MP has considerable legal ammunition to punish governmental agencies. For that reason, Lutti explains that, in most cases, he is able to settle the problem by negotiating with the accused party directly, without resorting to legal action. When we talked in 2013, his office was receiving roughly one noise complaint every day.

The Appellate Judge

Once one of the parties (or both) appeal to change a final sentence in the trial court, the case moves from one of the 319 judicial districts in the state of São Paulo to the Palace of Justice in downtown São Paulo. There is a range of possible ways to appeal to a higher court. One of the most common ones is what we will call "provisional appeal."[16] The main reason people resort to this instrument is to reverse a provisional sentence that is likely to cause to the party serious injury of difficult

---

[14] They are public because they involve the MP, and unconditional because the MP has the obligation to consider the case regardless of the victim's wishes.
[15] The MP can only ask for the case to be shelved if (1) it clearly does not constitute a crime; (2) the possibility of punishment has expired; or (3) the plaintiff does not have a legitimate role in the cause.
[16] *Agravo de Instrumento* in Civil Law and *Recurso em Sentido Estrito* in Criminal Law.

reparation. A successful provisional appeal needs to provide an unambiguous demonstration of both the likelihood of the allegations (*fumus boni iuris*) and the danger of waiting until the end of the proceedings (*periculum in mora*). However, it should not risk being irreversible, as it is provisional. The other common type of legal action is the appeal against a final sentence, which I will call a "final appeal."

Most of the Court of Justice chambers are formed by five judges, one of which acts as the president of the chamber. During the hearings, the chamber president sits at a large table above the other four judges. On his/her right side sits the state appeals prosecutor, who represents the Public Prosecutor's Office in Appeals Court and has the autonomy to agree or disagree with the state prosecutor. On the president's left side sits the chamber clerk, who checks documents, interacts with the lawyers, and makes sure that the hearings move smoothly. The chambers decide in the form of *acordãos*, or sentences based on "accords" on a panel of (at least) three judges. One of the judges on the panel is the rapporteur of the case. She is accountable for analyzing the case and proposing a sentence in a written document that follows the same tripartite structure mentioned above (summary-reasoning-verdict). Once the president of the (Public, Civil, or Criminal) section distributes the case to a rapporteur, the second and third judges are decided based on descending order of seniority. Most panel decisions are unanimous, with all judges following the rapporteur's position.

Three main events take place during a chamber hearing: preferences, oral arguments, and divergences. On the day scheduled for the session, the lawyers who want to hear the judges decide their case arrive early and ask the secretary to add their case to the "preferences" list. That way, they not only find out about the decision right away but also get to hear the reasoning behind it from the discussion between the judges on the panel (as opposed to the United States, this deliberation is open to the public). Lawyers who want to argue their case before the judges include their names in the "oral arguments" list. They patiently wait until the chamber president asks them to come to the lectern to "sustain" their arguments in a fifteen-minute oral argumentation (see Figure 5.2).[17] After the preferences and the oral arguments, the judges move to the divergences. As its name indicates, these are cases where there is disagreement between the judges on a panel.

Appellate judge Ricardo Torres de Carvalho, whom I interviewed in his office, not far from the Court of Justice, works in the 10th Public Law Chamber, which holds hearings every Monday. Besides his work in this chamber, Carvalho has

---

[17] As opposed to hearings in the United States, each party does not notify the other if it will take part in the oral arguments.

FIGURE 5.2. Inside the Palace of Justice: lawyer (standing on the left) at the 1st Environmental Chamber during the oral arguments.
Photo by the author, January 2017.

volunteered to work in the 1st Environmental Chamber (1a *Câmara Reservada ao Meio Ambiente*), which holds hearings every three weeks. Every week, the president of the Public Law section distributes the cases to all eighteen chambers and the two Environmental Chambers. On average, 300 of these go to de Carvalho's 10th Public Law Chamber, and 250 to the 1st Environmental Chamber. Every week, the 10th Chamber assigns to Carvalho about 30 final appeal cases to adjudicate as rapporteur; the 1st Environmental Chamber assigns him to roughly five final appeal cases. Once the appeals get to his office, the judge's secretary distributes the cases to Carvalho's four assistants. The assistants analyze and summarize the cases before handing them to judge, who will go over each one. He then submits the sentences that he prepared as a rapporteur to the other judges on his panel.

Besides these final appeals, de Carvalho also receives provisional appeals—roughly eight from the 10th Public Law Chamber and one or two from the 1st Environmental Chamber per week. We already know that these appeals entail pressing matters, which is why the appellate judge can receive them on his computer at any time during the day and issue a provisional sentence promptly. This decision is temporary, as it still needs to be decided with the entire chamber panel during the hearing. Whether we are talking about final or provisional appeals, the rapporteur is the only justice who goes over the entire case. However, the other judges on the panel can withdraw the case to analyze it more carefully at any point during the proceedings.

So where does an appeal related to noise, brought by lawyer Carneiro or state prosecutor Lutti, go once it arrives at the Appeals Court in downtown São Paulo? If there are many legal channels that lead from problem to legal action, there are also many places for it to be adjudicated. "We discussed this problem of noise a lot, especially the issue of competency. Nobody knows exactly where it belongs," admits Carvalho (personal communication, June 2017). But isn't noise pollution an environmental threat? And if so, wouldn't it belong to the Environmental Chamber? As the judge puts it, "The Environmental Chamber has more to do with what we refer to as the 'natural environment'" (personal communication, June 2017). Indeed, several accords I came across were not even discussed by the judges because of conflicts of competence—Public Law Chambers arguing that any noise issue should go to the Environmental Chambers, and Environmental Chambers sending appeals against municipal fines to the Public Law Chambers.

Jurisprudence

We now have a better sense of how lawyers, state prosecutors, and appellate judges convert noise litigation into governmental policing and punishment measures. We still need to identify: 1) what types of legislation, evidence, and arguments the parties deploy to revert provisional and final sentences, and 2) how the appellate judges tend to respond to these strategies. I focus on accords issued by the São Paulo Appeals Court between 2007 and 2017. The number of accords examined varies for each of the five types of litigation considered. In total, I examined roughly 500 accords: 130 related to the PSIU, 140 about private neighborhood law, 60 public civil action cases, 120 misdemeanors, and 20 environmental crime cases. My analysis focused only on provisional and final appeals.

FIGHTING THE PSIU

The previous chapter followed the administrative flows inside the PSIU initiated by Ms. Freire against the Bar da Esquina. We saw that the bar owner was convicted and ordered to pay a fine after the city rejected his administrative appeal. Let's continue following this case and see how the owner appealed to the Court of Justice to reverse the city's decision. The bar owner's lawyer uses different strategies to try to annul this administrative fine, but this is no easy task. That the fine was maintained after passing through such flows sends a strong signal to the judges about its materiality. As we already saw, the city has built several layers of checks and balances to protect itself from litigation from its citizens. In the vast majority of cases, the city's attorney general, with the assistance of the PSIU's juridical sector, succeeds in showing the appellate judges that there were no gaps in the administrative flows.

One thing the appellant can do is reuse the same arguments raised in the administrative appeal. The lawyer can call into question the PSIU's inspection procedures, arguing that the law requires the PSIU agent to conduct the noise measurements publicly with the bar owner present. The case rapporteur, however, explains that this is a misreading of the ordinance. The law requires the measurement *results* to be public, not the measurement itself—otherwise the bar owner would obviously stop the noise when seeing the inspector measuring it. The rapporteur also explains that the public official is legally imbued with "public trust" and the "presumption of legality," in which case the burden of the proof lays on the bar's side.

Another possibility for the lawyer is to argue that there was "defense restriction" (*cerceamento de defesa*). She explains that the city undercut her client's rights of defense (*direito de defesa*) by refusing to hear more witnesses. This strategy usually fails as well. The rapporteur reminds the lawyer that the city had enough evidence to make the decision. It is the judge's prerogative to evaluate whether there is enough information to adjudicate; that includes dismissing the collection of testimonial material. With few other options, the lawyer then argues that *the fine was abusive* because it can bankrupt the bar. After all, the Constitution asserts that the state must follow the principle of reasonability when confiscating private property and cannot lead to the total loss of the taxpayer's estate. Again, the rapporteur could explain that this is a misreading of the law. This confiscation veto does not apply to fines, only to taxes. The judge explains that the high value of the fine was reasonable due to the necessity of prohibiting disturbances of the peace in a way that both poor and upscale venues feel the heavy hand of the state.

Let's suppose Bar da Esquina broke the 1:00 a.m. Law rather than the Noise Law by remaining open with open doors until 2:30 a.m. The lawyer could then raise the issue of *business classification*, arguing that the 1:00 a.m. Law does not apply here because her client's venue is not a bar. Taking advantage of the blurry definition of "bar" we saw in Chapter 3, the lawyer explains that Bar da Esquina is a restaurant-cum-bakery-cum-bar, and for that reason is within the legal boundaries of Law 13772/2004 (which protects restaurants)[18] instead of the 1:00 a.m. Law. In this scenario, the lawyer reasons, only the zoning laws should be used to stipulate noise limits as Bar da Esquina does not have the time constraints that bars have. In response, the city attorney general uses a range of mechanisms to prove that the Bar da Esquina is, in fact, a bar. He points to the name of the venue (it has the word

---

[18] See Chapter 3.

"bar" in it!), includes a copy of the menu showing that the venue has a large collection of drinks typical of a bar, draws on the PSIU agent's testimony that the place did look like a bar, and indicates that, bar or not, the venue has a record of noise complaints and therefore needs to be punished.

To the distress of the bar owner, all strategies are likely doomed to failure. Argument after argument, the city's attorney general and rapporteur turn down attempts to annul the fine. One infrequent situation, however, does seem to be able to break the links between noise and fine. It occurs when the city fails to execute the fine in five years[19] due to some clogging in the administrative flows. In such cases, the judge has no option but to waive the punishment. As one appellate judge, Arthur del Guércio, explains in one accord, "the Administration is not exempt from following the actions it proposes, that is, moving the process forward and not abandoning its execution, running the risk of letting it expire."[20]

PRIVATE LAW

One of Carneiro's main tools is the 2002 Civil Code. Chapter 5 of the code addresses neighborhood law and includes a section on the abnormal use of property:

> Article 1277. The owner of a building has the right to eliminate interferences that are harmful to the safety, peace, and health of its inhabitants caused by the use of neighboring property.
>
> Single paragraph. Interferences are prohibited considering the nature of the use and the location of the building, according to the norms that distribute the buildings into zones and the tolerance limits of the residents.
>
> Article 1278. The right referred to in the preceding article does not prevail when interference is justified by the public interest, in which case the owner causing them shall pay indisputable compensation to the neighbor.
>
> Article 1279. Even if by judicial decision the interferences must be tolerated, the neighbor may require their reduction or elimination whenever possible.

As the middle ground between the more abstract constitution and local laws, the civil code offers general guides but requires some level of interpretation. Carneiro dedicates a good chunk of his book on noise litigation to unpacking

---

[19] Decree 20910/1932, Article 1, establishes that "The passive debts of the Union, the States and the Municipalities, as well as any right or action against the federal, state or municipal Treasury, of whatever nature, shall expire in five years from the date of the act or fact from which they originate" http://www.planalto.gov.br/ccivil_03/decreto/1930-1949/D20910.htm.

[20] Rapporteur Arthur del Guércio, Vote 19256 of Appeal 0155777-03.2008.26.0000 (January 17, 2013).

and interpreting those three articles from the code. How does one determine "abnormal use," "neighborhood," or "zones"? Regarding the first term, Carneiro considers it problematic because "normal" and "abnormal" refer to property use whereas "harmful" relate to the *effects* of property use. This, of course, has important legal ramifications: normality focuses on the individual making the noise, whereas harm focuses on those subjected to it (Carneiro 2014).

Following this rationale, Carneiro explains that a neighborhood is not determined by the distance between buildings, but by the web of actions to which a community is exposed. Like nineteenth-century communities in Europe organized according to the acoustic reach of church bells (Corbin 1998), Carneiro argues that sound determines the community rather than the other way around. This includes not only residents living blocks apart, but closer together (roommates sharing the same property). Finally, rather than simply defining the neighborhood according to the zoning laws, judges often prefer to rely on photos and the experts' assessment to get a sense of the characteristics of a specific neighborhood—in fact, the appellate judge's characterization of a zone may prevail over municipal zoning laws.

The first challenge for someone suing over noise using the private law channel is to show that there is a misuse of property. As legal scholar Maria Helena Diniz explains, this should be determined "considering the circumstances of each case, verifying the degree of tolerability, invoking the local uses and costumes, and examining the nature of the nuisance and how the property was used in the past" (Diniz 1995, 212). The judges do not consider a noise complaint legally valid if they think the nuisance is within the limits of tolerability for that place, taking into account the "average" resident. In this sense, they navigate between conventions about "normality" and "harm" to avoid basing their decisions on abusive noisemakers as well as on people extremely sensitive to noise.

The bottom line, the judges argue, is that collective life requires a certain amount of sacrifice, and we all need to adapt to minor annoyances, with the risk of clogging up the judiciary. This middle ground helps the judiciary navigate between the rest and health delineated by Ear 1.0 and the property rights that embody the foundation of private law. Appellate judges rely on the three concepts mentioned in Article 1277: safety, peace, and health. If the noise has a negative effect on anyone, then the judges will likely decide in favor of the complainant. Of the three concepts, the most certain path is health, which has already been paved by Ear 1.0, Ear 2.0, and the NBRs discussed in Chapter 2.

This leads us to another common controversy: the collection and assessment of the evidence. When a case reaches the court of appeals (especially provisional appeals), the judges often request that experts provide a reliable report, following

NBR 10151. The technical report from an expert is usually the most important piece of evidence here because, as one judge puts it, "it does not admit value judgment."[21] However, many rapporteurs remind the parties that other elements should be considered as well. Although judges tend to be suspicious of witnesses due to potential bias, they sometimes dismiss the expert's assessment when it comes to establishing an unambiguous causal chain between noise and harm. The experts often recognize the margin of error of the measurements, especially because they tend to be measured *after* the event that triggered the legal action has passed. Things can get more complicated when the measured values and the NBR 10151 values are close, or when two experts reach conflicting conclusions. In these cases, the judge will likely consider the witnesses and other pre-existing information.

Another common misunderstanding among the parties relates to the constitutional autonomy of the judicial spheres. First, there is confusion between public and private law. The judges often have to explain that even if the noise does not exceed the limits established by the noise ordinance (public law), one can still sue someone using the private law channel. Second, there is some lack of understanding about the difference between civil and criminal law. If the complainant goes through criminal law and the judge finds the defendant innocent, that does not mean the person is automatically acquitted in the civil sphere. Public, civil, and criminal litigation can inform but not determine the final rulings of one another.

Civil law rulings revolve around two premises: the obligation to do (and not to do), and material and/or moral damage compensation. A judge might order a wrongdoer to cease making noise under the penalty of a daily fine and to compensate the plaintiff for material and moral damages. There are plenty of controversies around moral damage compensations. When are they justifiable? What kind of proof is necessary? What is the appropriate amount? Doutrinador Sérgio Cavalieri Filho explains that "only pain, shame, suffering, or humiliation should be considered moral damages, which, departing from normality, interferes intensely in the psychological behavior of the individual" (Cavalieri Filho 2004, 98). However, the doutrinador adds, "mere unpleasantness, annoyance, irritation, and exaggerated sensitivity are outside the orbit of moral damage" (Cavalieri Filho 2004, 98). To evaluate moral damages, the judges consider the severity of the illicit conduct, the intensity and duration of the suffering experienced by the victim, the offender's financial situation, and the victim's personal circumstances.

---

[21] Rapporteur Roberto Maia, Vote 6979 of Appeal 9096120-74.2008.8.26.0000 (February 4, 2014).

As it involves litigation between two private parties, with the state serving simply as a neutral arbiter, this channel is particularly unstable and difficult for the noise complainant to move forward. Tensions between the specificity of the case and the generality of the law might be substantial. For that reason, the time frame for the case resolution and punishment can vary more widely. Besides, as Carneiro suggests in his book, in comparison to the other litigious channels analyzed here, judges might not feel the same pressure to create a more cohesive jurisprudence on neighborhood litigation, which makes the work of doutrinadores such as Carneiro (who is mentioned in many of the sentences I analyzed) particularly crucial.

PUBLIC CIVIL ACTION

The Public Civil Action Law[22] gives the MP the authority to propose legal actions for protecting public property, including the environment. As we saw already, the MP is part of the executive branch but has considerable autonomy in defending the collective interest and investigating both private and public institutions. More than in any other channel, the MP's cases are particularly complex and wide-ranging, and can take a range of actors to the court. It can do that not only because of its institutional status but also because it usually relies on a substantive body of evidence.

Like we saw in private law in relation to abnormal and harmful use, the appellate judges often do not consider the use itself, but rather the effects of the activities occurring on a given property as sufficient to characterize a deviation from its primary function. The MP's main targets include large venues such as social clubs and soccer stadiums. Other targets of the MP are smaller venues, including churches, restaurants, and bars. In the period examined (2007–2017), the major events relate mostly to music concerts at sports arenas, which the MP sees as secondary to the venues' social function. This rationale also allows the judiciary to break legal walls separating juridical and natural persons. In qualifying the activity as *abusive use* of property, the MP can reach the club owners' assets.[23] The MP can request that the city stop issuing licenses

---

[22] Federal Law 7347/1985.
[23] Art. 50 of the Civil Code (2002) states that "in the case of the abuse of juridical person, characterized by finality deviation [. . .], the judge may decide, at the request of the party or of the Public Prosecutor's Office, when it may intervene in the proceedings, that the effects of certain obligation relations are extended to the private assets of the legal entity's managers or partners."

for musical activities in these venues. Appellate judges prefer to err on the side of the environment (*in dubio pro ambiente*), taking into account the 1988 Constitution's statement that both current and future generations have the right to an "ecologically balanced environment" (Art. 225). However, unless it has become clear that music events at the venue will likely exceed the noise limits, appellate judges consider preventing the event too drastic, preferring to allow the city to issue licenses but requesting the monitoring of the noise levels generated by the event.

The city can attempt to detach itself from the MP's legal action. In provisional appeals, the city's attorney general often deploys the concept of "passive illegitimacy," claiming that the MP was not able to prove the nexus of causality linking the fact with the city itself. The MP tries to persuade the judges that the nexus exists by claiming the state agencies have neglected their obligations. The city's attorney general then maintains that the city has "administrative discretion." As legal scholar Hely Lopes Meirelles explains, this is the prerogative "conferred on the public administration, explicitly or implicitly, for the practice of administrative acts with the freedom of choosing their convenience, opportunity, and content" (Meirelles 2016, 139). The MP then responds by distinguishing discretionary power from administrative inertia. The former entails the authority to decide how and where to enforce the law, the latter the failure to act when the law has been broken.

Let us continue to follow the case described in the previous section. We saw that in a provisional appeal, the MP asked the city to revoke the bar's permit and close it down, and the state to deploy its police force to patrol the neighborhood. It also demanded that the PSIU change its operations to stay accessible 24/7. The case rapporteur orders the immediate cessation of any noise above the NBR 10151 limits but says he cannot close the bar at that point in the proceedings without an impartial technical report. In the final sentence, after a technical report states that the bar had indeed produced excessive noise, the panel establishes a fine for every day of the infraction. Despite attempts by the city to assert its administrative discretion, the panel concurs with the MP that the city was negligent, as it was aware of the infraction but failed to act. The court then fines the city administration, saying that it should seek compensation from the responsible public officer so that the population does not have to pay for the neglect of the administration's agents.

Regarding the collective moral damages compensation requested by the MP, the panel explains that environmental damage does not cause moral damages to the community unless the action has generated "a feeling of social unrest, uneasiness, and chagrin" (quoted in Delgado 2008, 105). In none of the cases I examined did the MP succeed in getting moral damage compensation for the community affected by the noise.

MISDEMEANOR

The most frequent culprits here are quarrels between family members and neighbors, as well as loud music from churches and bars. In Chapter 4, we followed the police administrative flows and Major Tenório's frustration with noise complaints not turning into police reports. The complaints that do succeed in becoming police reports, evolve into legal action after passing through the desks of the civil police commissioner, judge, and state prosecutor. Article 42 from the 1941 Misdemeanor Law, mentioned in Chapter 3, establishes as an infraction:

> Disturbing someone's peace or work by (I) shouting or causing uproar; (II) conducting noisy or annoying activity in contravention of legal prescriptions; (III) using sonic instruments or acoustic signals abusively; (IV) provoking or not preventing noise made by animals under one's guardianship.
> Punishment: simple prison[24] from 15 days to three months, or a fine.

One common appeal comes from the MP when trying to reverse a trial court decision to shelve the case. This happens when the judges are not certain how many people were affected by the disturbance. There is enough jurisprudence supporting the view that Article 42 should be used only when the noise disturbs several people. Similar to what we saw in private law, the judges usually try to reach a middle ground between rest and leisure, and between individual and collective rights by focusing on the "average" citizen. If only one person complained, they conclude, it is more likely that that person's sensitivity is just above average and that there is no clear path for litigation.

The second controversy relates to what type of evidence counts. As one judge explained, the move from the executive to the judiciary branches marks a shift from erring on the side of the victim (*in dubio pro societate*) to erring on the side of the accused (*in dubio pro reo*). Unlike private law, the materiality of the case here relies to a greater extent on oral testimonies from the victim, defendant, and witnesses. A common defense strategy is to claim that oral testimonies are biased and, for that reason, the MP does not have enough evidence. Although most appellate judges maintain that Article 42 does not require measurements by an expert,

---

[24] In a simple prison punishment, the person convicted works and/or sleeps in a minimum-security facility.

they might see the expert report as the only alternative to prove that any harm was done if there are not enough witnesses.

Finally, there are controversies about proper punishment. Convictions in the criminal sphere are calculated following three steps. First, the judge considers basic facts such as the offender's degree of guilt, background, social conduct, and motivation, as well as the circumstances and consequences of the crime. Having established this base sentence, the judge then focuses on the mitigating and aggravating factors (e.g., the offender's age, lack of knowledge of the law, confession, recidivism, vain motive, or attempts to cover up crimes). The final and third step relates to special causes (e.g., use of a firearm or participation in a criminal organization), which can increase or decrease the sentence between one-sixth and one-third of the time.

The judges tend to prefer fines to prison time. For punishments of one year or less, the judge can replace the sentence with a fine or community service. As legal scholar Ataliba Nogueira notes, "In short-term sentences, imprisonment is avoided under the argument that it only entails serious inconveniences to the convicted by removing him from his occupation and staining his life with the shame of imprisonment, not to mention exposing him to the corruption that is particular of prisons" (quoted in Costa Leite 1962, 31). Even when the offense combines Article 42 and other more serious offenses, it is common to see appellate judges trying to avoid confinement.[25] In one example, the defendant was convicted for disturbing his neighborhood by playing loud music in his house (Art. 42) and insulting the police officer who was responding to the complaint (contempt of police, Article 331 of the criminal code). The trial court sentenced the convicted to six months of prison (for Art. 331) and fifteen days of open prison (for Art. 42),[26] replacing the prison time for community service. The defendant appealed, claiming insufficient evidence. The appellate judges decided to replace the community service with a fine of two-thirds of the minimum wage.[27] This *in dubio pro reo* tendency in the Court of Justice explains the military police's frustration from the previous chapter. After all, this is their main instrument for disciplining noisy citizens. To their exasperation, once the offense moves to the judiciary branch, either the punishment is a few hundred dollars or less, or the case arrives at the judge's desk so late that the punishment is annulled.

---

[25] Other common offenses include threatening (Art. 140 of the Criminal Code) and insulting (Art. 147 of the Criminal Code) someone.

[26] PCL sentences do not include closed incarceration. Open prisons have low security level and often allow the convicted to sleep at home.

[27] Minimum wage in Brazil is calculated monthly. For reference, see https://tradingeconomics.com/brazil/minimum-wages.

ENVIRONMENTAL CRIME

This is the least stabilized channel and thus a risky move for the MP. As briefly discussed in Chapter 3, Article 54 from the 1998 Environmental Crime Law establishes as a crime "causing pollution of any kind at levels that result or may result in damage to human health, or cause the death of animals or significant destruction of flora." The punishment varies from one to four years of prison. We saw that the bill sent to the president included an article dealing specifically with noise pollution, but the president ended up vetoing it, arguing that Article 42 from the misdemeanor law already covered noise in the criminal sphere. That veto removed a clear path for interpreting noise as an environmental crime. That is, the appellate judges look at Article 54 from the Environmental Laws and then look at the vetoed Article 59 and assume that Article 54 was not intended for noise. As judge Carvalho already suggested, one common reading of Article 54 limits it to the "natural" environment, which includes flora and fauna only. When it comes to loud sounds (even *very* loud sounds), the judges often do not see enough evidence linking noise to health damage (*in dubio pro reo*).

The presidential veto also gave the judiciary an easier route to detach noise from the Environmental Crime Law and reattach it to the misdemeanor law (Article 42). This rechanneling makes a considerable difference. For example, in 2012 the MP sued a construction materials company and its three owners under Article 54, arguing that it generated noise pollution between 2009 and 2011 during weekdays, weekends, and holidays, day and night. An expert report showed that the sound pressure levels indeed exceeded the limits established by NBR 10151. The report was corroborated by police reports, oral testimonies, and photos. Even the defendants admitted that their activities were loud. The trial court judge sentenced the co-owners to one year of prison (replaced by community service) and two minimum wages' worth of fines. It also sentenced the company to pay the city US$5,500, to be transferred to environmental initiatives.

The appeals court, however, did not see enough evidence showing the company's noise *actually* damaged human health. What they saw, drawing on the witnesses' own use of the term "disturbance," was a disturbance of the peace. Once the crime category was rechanneled to a misdemeanor, the company was automatically off the hook, as juridical persons cannot respond for misdemeanors. The co-owners were sentenced to fifteen days of simple prison, which the appellate judges replaced with a fine worth up to one-third of a minimum wage. This illustrates the considerable difference in punishment when noise litigation moves from environmental crime (as pollution) to misdemeanor (as peace disturbance).

This does not mean there is no jurisprudence linking Article 54 to noise pollution. When the MP manages to collect several expert reports from different state agencies showing that sound pressure levels are well above the NBR 10151 limits, the judges feel more confident in claiming that there was noise pollution. In such cases, they make sure to emphasize the number of people affected and the reference to "pollution of any kind" mentioned in Article 54. To stabilize noise as a viable entity in this legal channel, the sound specialists still need to establish stronger links between environmental noise and health problems. This requires the continuous fine-tuning of Ears 1.0 and 2.0, which would allow doctors to state with precision what specific health conditions are harmed by x hours of noise exposure at x levels. How this could be done outside environmentally controlled factories and offices, however, is not clear. At this point, the São Paulo criminal law appellate judges are divided between those who follow well-established jurisprudence on peace disturbance (Article 42) and those (few) who see evidence linking noise and the Environmental Crime Law.

Multi-Channel Sound-Politics

In this chapter, we have considered yet another mode of existence that mediates attempts to stabilize and neutralize noise controversies. We examined the Court of Justice by visiting three legal specialists and saw how each of them shapes sound-politics in the judicial sphere by carefully tying together documents across legal channels to convert noise complaints into punishment. From the lawyer and the state prosecutor, we learned how to build a case. When trying to put an end to sonic disturbances, these specialists usually mix the channels considered here; they initiate legal actions in both civil and criminal law. As the cases move forward through multiple legal channels, they require so many resources from the defendants (with the torrent of evidence and the punishment attached to it piling up) that they often decide to move somewhere else even before the final sentence. As Carneiro explains, "The strategy is to kill an ant with a cannon, because sometimes this ant can be very strong" (personal communication, June 2017).

Finding the right way to move in each legal channel is not an easy task for the "average" lawyer, as showing proof in noise litigation is relatively complex, not only because of its techno-scientific peculiarities but because the issue permeates different legal fields. Carneiro prefers to stay away from sub-federal legislation as much as possible. "If there is a federal norm saying that the limit is 30 dB, the municipal government can't simply create a zoning law and establish the limit of 50 dB because this creates a legislative conflict. In such cases, the federal regulation should supersede the municipal one" (personal communication, June 2017). The reason for staying at the federal level is that NBR 10151, linked to CONAMA's regulations, provides lower decibel limits than most zoning laws. The judges tend to not pay much attention to the difference between specific noise and background noise, as indicated in NBR 10151. Instead, they compare the defendant's specific noise with the NBR 10151 decibel values table. That makes it very easy for the noise in question to be above the limits and very hard for the defendant to claim there is no disturbance of the peace. This is also crucial for understanding why the revisions of NBR 10151 examined in Chapter 2 have generated so many heated debates. The state prosecutors can include governmental institutions in their legal action. To do that, they examine municipal and state legislation and show that neither the city nor the state agencies have done their jobs properly. As a result, the MP concerns itself primarily with federal legislation, and mobilizes more localized laws to prove the executive branch is not doing enough to safeguard the common interest.

In the previous chapter, we saw that the nuisance/decibel conundrum was crucial for understanding sound-politics. I argued that the police and PSIU operate with different mechanisms of state power because the former acts in the name of peace and the latter in the name of noise pollution control. In this chapter, we saw that these two concepts are central here as well, depending on the litigious field considered, with each requiring a certain type of evidence (oral witnesses or expert report). Carneiro has mixed feelings toward the increasing importance of measurability in noise litigation. "On the one hand," he explains, "decibel limits establish a concrete parameter for something one could think is subjective. On the other hand, you create the impression of a false requirement: if it is below the limits, it can't disturb anyone" (personal communication, June 2017).

We now have a well-rounded perspective of sound-politics in São Paulo in the legislative, executive, and judiciary branches. We saw how chains of reference, political circles, administrative flows, and legal channels all affect the extent and intensity of state intervention in a given noise controversy. We have traced the debates that preceded and followed such controversies and identified specific actors involved in moving them toward a more stable state of affairs. Inside the Palace of Justice, we observed how judges, state prosecutors, and lawyers deploy legal documents to move a case forward and examined the most common strategies used by plaintiffs and defendants to get the state on their side. We are finally ready to leave (and later reenter) downtown São Paulo, where the municipal institutions are headquartered (as shown in Figure 5.1), to examine in more depth the noise coming from São Paulo's urban fringes.

> The person who haunts us is the person who is
> having more pleasure than us.
> ADAM PHILLIPS, *On Balance* (2010, 25)

6

THE "ROWDY" TEENAGERS

IN A 2012 article, Teresa Caldeira described how the city's lower-class suburban youth have been deploying a range of practices for recreating the public space (Caldeira 2012). Caldeira builds her argument around two groups: the *pixadores* and the *motoboys*. The former is involved with *pixação* (or "tagging"), a unique form of public inscription[1] that grew independently from graffiti scenes elsewhere and is commonly seen on buildings and monuments of all sizes and shapes in São Paulo. Since pixação emerged in the 1980s, pixadores have established a complex network of illicit visual signs across the city. The practice is a combination of individual bravado, radical sport, and turf affiliation (pixadores include their region in the tags). The cryptic aspect of these signs, created with a "2- or 3-inch foam roller with an industrial color of bucket paint" (Manco et al. 2005, 27), is distinct from the more legible images and political messages of other graffiti styles. Pixação has been a "vehicle for the youth of the city to assert their existence and self-worth, and to do it loudly" (Manco et al. 2005, 29).

Motoboys are service workers paid to move across the city on their motorbikes to deliver all sorts of goods, more often restaurant food, pharmacy medicine, and

---

[1] In São Paulo, graffiti and *pixação* are two different things. Graffiti is seen as art and has "become a type of relatively sanctioned public art in São Paulo and is so prevalent that it has become its own tourist attraction" (Caldeira 2012, 395). *Pixação*, on the other hand, is usually considered vandalism.

documents. Between 2001 and 2010, the number of motorbikes in São Paulo grew 118%. In the early 2010s, of the almost 900,000 motorcycles circulating in the city (12.7% of the total fleet), roughly 200,000 were used by motoboys.[2] With their zigzag riding style, motoboys cut across heavy traffic using the empty spaces between lanes, often riding closely between car side mirrors. Their presence is often felt from afar, as they honk constantly to warn car drivers not to change lanes. On weekdays, they are omnipresent in the city's main avenues, "physically, noisily, close by" (Caldeira 2012, 412). In a city where, in 2016 alone, the government issued on average twenty-nine traffic fines per minute,[3] car owners have expressed concern about what they see as the motoboys' "risky" conduct (Silva et al. 2008). In 2010, 37.5% of the people killed in traffic accidents in São Paulo were bikers.[4]

Caldeira's interest in these youth urban practices in post-dictatorship Brazil lies in their knack for evidencing "the limits of the democratization process by simultaneously expanding the openness of the democratic public sphere while challenging it with transgressive actions ranging from the mildly illicit to the criminal" (Caldeira 2012, 385). Indeed, important elements connecting these nonwhite male-dominated practices include the importance of moving around the city as strategies for turning public spaces into turf zones deeply embedded in subcultural values. These practices challenge the state and reframe notions of public and private spaces through relatively risky and (often) illicit behavior.

In this chapter, I focus on another marginalized youth group that since the mid-2000s has been reshuffling public spaces. Like the pixadores and the motoboys Caldeira describes, this group has been enmeshed in debates about leisure, civility, taste, and right to the city. But unlike than the pixadores' imprinting and the motoboys' zigzagging, which take place in more central spaces, the events examined here happen in the city's vast poor peripheries. I consider the spread of street parties known as "pancadões," or "big thumps," a term that alludes to the loudness of the music beat. In following the pancadão controversy, I am interested in understanding the investment of specific groups as they try to either maintain or eliminate this sound (music or noise?) from the streets. After locating music within sound-politics, I consider the actors necessary for mobilizing the party. These include sound equipment, cars, body stimulants (licit but often illicit

---

[2] "Motos em São Paulo crescem 118% na década e mudam trânsito da cidade," *UOL Notícias* (April 5, 2011), https://noticias.uol.com.br/cotidiano/2011/04/05/motos-sao-13-dos-veiculos-de-sao-paulo-e-exigem-mudancas-de-atitude-no-transito.jhtm.

[3] "Prefeitura de SP aplicou mais de um milhão de multas por més em 2016," *Folha de São Paulo* (April 4, 2017), http://www1.folha.uol.com.br/cotidiano/2017/04/1872578-prefeitura-de-sp-aplicou-mais-de-um-milhao-de-multas-por-mes-em-2016.shtml.

[4] "Capital registra 10 mortes envolvendo motos por semana," *O Estado de São Paulo* (May 3, 2012), http://sao-paulo.estadao.com.br/noticias/geral,capital-registra-10-mortes-envolvendo-motos-por-semana,868383.

as well), cell phones with Internet service, and resonant unattended streets. The pancadão has become a central event for the youth (most of them minors) to dance, consume alcohol, flirt, challenge the police, and strengthen friendship ties away from the surveillance of parents, teachers, and bosses. As I show, the explosion in popularity of funk carioca (to which the term "pancadão" is associated) and its local variant, *funk ostentação* ("ostentatious funk"), were significant for the proliferation of these parties in São Paulo's suburbs.

In the final section, we are going to consider some of the groups invested in disarticulating the pancadão in the name of public security, good taste, family values, civility, and right to rest. I show how the broken windows and zero tolerance approaches have influenced residents and public officials to hear the pancadão as potentially dangerous and criminal. We already came across Wilson and Kelling's article about the effects of the environment on crime, in which the authors claim that "one unrepaired broken window is a signal that no one cares, and so breaking more windows costs nothing" (Wilson and Kelling 1982). For the authors, unattended spaces stimulated the presence of unwelcomed "strangers" to the community, including drug dealers, prostitutes, robbers, panhandlers, the "mentally disturbed," and "rowdy teenagers" (Wilson and Kelling 1982). To understand the disarticulation of the pancadão, I draw on archival research and participant observation at the Community Security Council (CONSEG) meetings (introduced in Chapter 4) in São Paulo's suburbs, and on interviews with the police officers, lawmakers, and the PSIU agents. I also rely on participant observation from several anti-pancadão operations, organized by the police, which I have had the opportunity to accompany together with the PSIU agents.

There are a few reasons why I focus on the pancadão controversy—and why I do so at the end of the book. During fieldwork in 2012, this was widely circulated in the press as a new public nuisance. For that reason, its relative fresh insertion in the city's soundscape allows us to examine how the governmental institutions described throughout the book have tried to reshuffle its networks to neutralize the problem. I argue that the existent technical standards, noise ordinances, law-enforcement mechanisms, and jurisprudence all required some tweaks to deal with the particularities of the pancadão. The measuring procedures of NBR 10151, which, as we saw in Chapter 2, rely on Ears 1.0 and 2.0, are not ideally suited to handle the pancadão's powerful bass. In the early 2010s, the police and the PSIU tried to fight the pancadão by organizing a large operation. State prosecutors made sure the legal channels could be deployed to punish pancadão organizers, and lawmakers worked to pass ordinances better tailored to terminating these street parties. By focusing on this controversial sound, we can thus follow a cross-modal analysis and see how the networks described in the previous chapters can cooperate with each other.

One of the major challenges for city administrations in Brazil after two decades of military dictatorship has been to increase participation in decisionmaking procedures. The pressing question is how to incorporate the country's large youth population in policymaking and how to create zones of dialogue that allow them to participate in the management of their own leisure activities. The pancadão controversy revolves around the most disenfranchised group examined in this book. Unlike religious groups, bars owners, traffic and construction companies, public officials, and acousticians, the city's segregated teenagers have virtually no representatives defending their interests in the municipal institutions, quite the contrary. São Paulo's funk carioca youth are marginalized because of where they live, their race, their age, their class, their taste, their dressing style, and their use of space and language. As Herb Childress puts it, "Teenagers have limited ability to manipulate private property. They can't own it, can't modify it, can't rent it. They can only choose, occupy, and use the property of others. [. . .] They stand in strong opposition to modernist models of private property" (Childress 2004, 196). In that sense, the pancadão provides a unique opportunity for understanding sound-politics at the junctions of noise, youth, space use, and citizenship in contemporary Brazil.

Loud Grooves

The pancadão controversy requires us to take a closer look at the place of music within sound-politics. So far, the advantage of building the book around sound-politics was that we could focus less on the specific sound than on how it became a noise issue for the state. In their reading of ANT, Emilie Gomart and Antoine Hennion suggest that in the "world of strong sensations" active and passive modalities are often blurred (Gomart and Hennion 1999, 221). The authors argue that action, the domain of scientists and engineers, is too subject-oriented an approach for dealing with passion. What is action and who acts in a musical attachment? "Passion, emotion, being dazzled, elation, possession, and trance [. . .]," they claim, "describe a movement in which loss of control is accepted and prepared for" (Gomart and Hennion 1999, 226–227).

Musical sounds are attention vortexes in an ocean of city-noise. Firmly attached to taste and intentionality, they form ephemeral pockets of attachment, usually surrounded by vast zones of detachment and distaste. Following Gomart and Hennion, I suggest we approach this vortex without pre-establishing the set of actors called upon to mediate a musical experience. Rather than considering music as group affiliation or individual trajectories or embodied experience or music genre genealogies or sonic tasting or audio technologies, we will remain close to ANT and follow all these entities as relevant actors in the articulation of the pancadão.

Steven Feld discusses music grooves as patterns in motion and as "recurrent clustering of elements through time" (Feld 1988, 74) that are "instantly perceived, and often attended by pleasurable sensation ranging from arousal to relaxation" (Feld 1988, 75). Musical grooves are easily identifiable, transposable, and attachable. Charles Keil defines "participatory discrepancies" the "personally involving and socially valuable" tasting of grooves, which operates collectively through minuscule embodiment differences during music experiencing (Keil 1987, 275). The discrepancies that the grooves stimulate are often performed via foot tapping, hand clapping, stomping, arm swinging, pelvis shaking, head nodding, and other collective choreographies of groove-tasting.

To get into the groove is to establish an affective attachment. For Feld, music styles operate as "cultural grooves," inscribing collaborative expectancies in time (Feld 1988, 74). The author argues that "grooves/styles are universes of discourse, pervasive, rigorous unities, assertions of control, and algorithms of the heart essential to affecting presences" (Feld 1988, 107). Grooves in songs (the beat) and grooves as music styles (as a relation between songs that creates time expectancies) are devices that mobilize listeners by translating a set of body-affective instantiations—what exactly these instantiations and attachments are can only be determined with ethnographic work. The sound-politics of popular music composes spaces and bodies through the double movement of the groove: as a musical recurrence and as a style. The repetition and difference that this double propagation entails operate particularly well to generate acoustic vortexes in the city. This makes it difficult for those *not* interested in the grooves to escape from their attention-grabbing quality.

To approach the pancadão's sonic dominance as a matter of groove attachment is only part of the story. The other part is its loudness. In our investigation of the Ear 1.0 in Chapter 2, we saw that standardization of an average ear was made possible with the establishment of hearing thresholds across the frequency/amplitude/exposure nexus. Once you bring passion to the equation, however, bodies can bend the norms. Some authors have argued that the ear responds differently to loud music in comparison to other sounds at the same decibels due to the working of a specific fluid-filled sac in the inner ear (Dibble 1995, Todd and Cody 2000). The sac (also known as the "primitive ear") works in conjunction with the cochlea, a more "sophisticated" sound detector and analyzer. Besides helping the brain interpret head movements through the vestibular system, the sac has optimal auditory sensitivity at lower frequencies. According to Neil Todd, the "human compulsion to exposure to loud, low-frequency sounds is a kind of acoustic equivalent of vestibular self-stimulant" (Todd 2001, 381). As the sac requires relatively high doses of sound energy for maximal stimulus, the cochlea needs to readapt to the loudness of the environment. As Todd explains, "by reducing the input gain of the cochlea, the balance between any pleasant saccular sensation and the unpleasant cochlear sensation will be swung in favor of the former" (Todd 2001, 386). This suggests the human ear's resilience in being black-boxed and its potential for destabilizing Ears 1.0 and 2.0.

A better understanding of Ear 1.0 has allowed music producers to explore loudness in music production. David Novak explains that "a louder sound is 'flatter' across the frequency spectrum and will be perceived as closer, fuller, deeper, and 'brighter' (more high frequencies). Because 'present'-sounding recordings are equalized to mimic these changes in perception of loudness, they can seem louder than they really are, even at low volumes" (Novak 2013, 54). Music genres tend to build forms of groove tasting around loudness in particular ways. Writing on the engineering of heavy metal sound, Robert Walser argues that "Intense volume abolishes the boundaries between oneself and such representations; the music is felt within as much as without, and the body is seemingly hailed directly, subjectivity responding to the empowerment of the body rather than the other way around" (Walser 2014, 45).

With increasingly sophisticated sound compression technology to push the sound closer to the passionate ear, loud grooves have become increasingly palpable. Julian Henriques's discussion of the heavy-bass grooves of the Jamaican sound systems shows how artists and MCs "plug their ears" into the speakers to come up with more graspable and powerful grooves, around which competition between DJs revolves (Henriques 2008, 19). "For the crowd out for the night," the author explains, "One of the distinctive features of a dance hall session is certainly the loud volume of the music [. . .]. The crowd experiences such cellular intensities as the sheer immersive weight, liminal force, and substantive presence of the sounding—impossible to escape or deny" (Henriques 2008, 53).[5]

---

[5] The Jamaican sound system culture was a major site of experimentation with sub-bass in music both in terms of production and performance. Dub, the music style born in the 1960s in homemade studios in Jamaica, explored the bass lines in reggae songs by remixing them and adding effects. Sound systems were mobile and played recorded music at parties both indoors and outdoors. New York hip-hop incorporated Jamaican bass materialism and the sound system apparatus. Kool Herc, a Jamaican migrant and pioneer in New York hip-hop scene, explains that he had a sound system "more powerful than the

Hip-hop has been another laboratory for experimenting with loud grooves in the lower frequency. As studio engineer Steve Ett explains, "Rap is a matter of pumping the shit out of the low end. The bass drum is the loudest thing on the record" (quoted in Rose 1994, 76). Tricia Rose argues that the loudness of heavy-bass grooves is also a matter of professionalism, as the bass drum and sub-bass notes should not completely lose definition at higher decibels (Rose 1994, 75). In 1980s Miami, a subgenre of hip-hop known as Miami bass gained national projection.[6] The music included a powerful resounding bass drum sound, a party-oriented vibe, lyrics permeated by call-and-response, suggestive dance gestures, and sex-related humor. Luther Campbell, the leader of Miami bass group 2 Live Crew, used to slow down Jamaican dub records at parties to make the bass notes boom extremely low. As Luke recalls, "At first you needed sixteen speakers to be competitive, then it went to twenty-four, then forty-eight, then sixty-four" (quoted in Westhoff 2011, 24). Bass power gained a new magnitude with the release of the Roland TR-808, one of the first programmable drum machines—ridiculed by professional studio engineers for its "cheap" sounds. As 2 Live Crew DJ Mr. Mixx explains, "There's a bass knob that if you turn all the way to the right it makes it seems like the bass is oversaturated. Back then I wondered why [other DJs] didn't make the records with the knob turned all the way around. I guess they thought it was too much bass" (DJ Mr. Mixx 2013). Black diasporic music styles such as reggae, dub, and hip-hop have established groove tasting as the tactile quality of low frequencies, what Steve Goodman refers to as "bass materialism" (Goodman 2010). In this music genres, sound engineering practices operate through vibrational modulation and take part in a kind of sonic warfare, "In a spiraling logic of hype escalation, intensification, and mobilization of the dance" (Goodman 2010, 28).

Funk Carioca and Funk Ostentação

In 1989, Rio de Janeiro-based DJ Marlboro released *Funk Brasil Vol. 1*. The album included rapping and singing in Portuguese over the instrumental Miami bass tracks. It sold 250,000 records and helped to consolidate a new genre of Brazilian electronic music known as *funk carioca* (funk from Rio de Janeiro).[7] Drawing on Miami bass, funk carioca has close ties with sexual content and humor, featuring a vocal style that combines speaking, shouting, and singing, with songs that tend

---

average DJ's speakers and surprisingly free of distortions, even when played outdoors" (quoted in Rose 1994, 52).

[6] For a discussion of Miami bass history, see Sarig (2007).
[7] For a discussion of the genealogy of funk carioca, see Cardoso 2018b.

to be concise and targeted to the dance floor. Funk carioca quickly became a discursive matrix about everyday life in the urban margins across the country. In the 1990s, sex-oriented and crime-oriented songs became two popular variants at funk carioca parties. In 1993, *O Estado de São Paulo* introduced "pancadão" as a term used by the funk carioca community in reference to the beat and to the music genre.

Funk carioca was relatively scarce in São Paulo until the mid-2000s, when funk parties in low-income suburbs started to grow. But in São Paulo a new spin was applied to funk carioca, along with a new label: *funk ostentação*. Here, personality and self-presentation are key and so is the "ostentatious" display of conspicuous consumption.[8] In 2008, MCs Backdi and Bio G3 released the song "Bonde da Juju" ("The Juju Crew")—"Juju" refers to the Juliet sunglasses series made by Oakley, retailing from $350 to $600. Besides Juliet sunglasses, the song also name checks 18-karat gold necklaces, Marc Ecko watches, and the Nike Shox series. Likely due to the over-saturation of the city's more politicized rap,[9] shifts in generational taste, the consumer boom of the 2000s, and increased demand for live music, funk ostentação struck a chord.

From this point forward, local MCs would build their fan base around songs celebrating access to cars, motorcycles, whiskey, and Cuban cigars, releasing increasingly ostentatious YouTube music videos. Konrad Dantas (aka Kondzilla), the funk ostentação music video director who helped to popularize the subgenre, defines the visual language of his music videos as "visual pleonasm": MCs show in the music video all objects of desire mentioned in the lyrics (personal communication, December 2012). Soon enough MCs started to include in their lyrics the ostentatious objects they would be able to show in the music video. In 2012, as it went unnoticed by record labels, TV networks, and radio stations, funk ostentação made the top ten music video list on YouTube with four songs. As of September 2017, Kondzilla's YouTube channel had almost 18 million YouTube subscribers.

São Paulo's sizable youth population assimilated funk carioca through the coastal cities of Santos and São Vicente. Funk ostentação first exploded in Cidade Tiradentes, a lower-class subprefecture in easternmost São Paulo, when local MCs, DJs, and producers decided to collaborate with the subprefecture to promote a funk carioca free of explicit and violent content. Deliberately avoiding contentious content such as the sex-saturated and the crime-oriented "gangsta" funk carioca subgenres was a strategy for entering São Paulo's vast and lucrative nightclub

---

[8] A cognate for "ostentatious funk," in terms familiar to North American hip-hop, the new sub-genre falls somewhere between luxury rap and swag rap.
[9] See Cardoso 2018b.

network (including in upscale districts). In 2009, with the help of Renato Barreiros, a cultural articulator and at the time subprefect of Cidade Tiradentes, the first Funk Festival gathered roughly 30,000 youth from all over São Paulo. Beyond the localized attempt from the Cidade Tiradentes subprefecture to legitimize the funk ostentação movement, and as the genre continued to gain popularity, unregulated street parties spread in the city peripheries. In 2011, the press started calling these unsupervised and spontaneous street funk parties "pancadão."

Articulating the Pancadão

During fieldwork, I identified three main versions of pancadão. The first variant, what I call the "mini-pancadão," relies on a small number of participants and requires very little planning. It involves youth from the neighborhood hanging out on the street, listening to music and drinking around the car. It can happen during the day or night and include other musical styles besides funk carioca. These hangouts generate noise complaints and tension between neighbors and are particularly common during weekends. The mini-pancadões take place next to one of the participants' houses, as this excerpt from Alexandre Pereira's ethnography in Brasilândia, a district on the north side of São Paulo, illustrates:

> I observed regularly [. . .] a pancadão on one of the streets carried out by young residents who stayed on the sidewalk listening to music at maximum volume. The street was connected to a series of alleys that led to small areas with a dozen houses. Due to the inexistence of backyards and the small space in the houses, most of the residents used the street as a space of coexistence. [. . .] At one point one of the residents called the police, who asked the father of one of the youths to turn down the volume. The youths turned it down momentarily but then turned it up even louder when the police left, while the father scolded the potential whistleblower. (Pereira 2010, 62)

The second type of event is what I call the "staged pancadão." Larger than the first type, it was initially organized as an extension of the *quermesses*. The quermesses are events that celebrate Saint John the Baptist, when the streets in several Brazilian cities become a space for the community to get together, play games, eat popcorn and fried cassava, and drink *quentão*.[10] In the suburbs of São Paulo, in the winter months of June and July, it is common for local churches and

---

[10] A hot drink made of wine, sugar, ginger, clove, and cinnamon.

other organizations to sponsor well-attended quermesses on their property with live music and traditional games. Due to the number of people circulating, the city subprefectures often block various streets from vehicles. The quermesses bring hundreds of teenagers to the streets and establish an atmosphere of festivity in the neighborhood.

As funk carioca spread in the suburbs of São Paulo, youth groups started to use the stages mounted for the quermesses for funk carioca and funk ostentação concerts. Due to the carnivalesque and spontaneous quality of the quermesses and the community's higher tolerance for loud music, the state initially did not interfere with the staged pancadões. Market stalls selling food and beverages welcomed the expansion of the party and soon informal street vendors started to congregate as well. The parties extended across suburbs, strengthening inter-neighborhood ties and articulating a space for youth-oriented musical performances. In 2011, anthropologist Gilberto Moreno attended a staged pancadão in Capão Redondo, a peripheral district on the south side. His account gives us an idea of what happens during these parties:

> Arriving at the place I saw a group of guys, in a dark alley, leaning against the unfinished brickwork of the houses. Walking 30 feet I could already hear the funk beat [. . .]. Market stalls sell drinks and foods and many youths circulate. The street is very narrow, giving the impression that there is a large crowd. [. . .] The *pancadão* is further down. Drug commerce and consumption is intense, from alcohol at the tents (at the entrance) to ecstasy, cocaine, and marijuana, sold around the party and in alleys. At end of the street, there is a stage set up especially for the party. On the stage, an MC introduces the attraction for the night [. . .]. (Moreno 2011)

As the events started to take place on a regular basis and gather more people, the neighbors' tolerance decreased, and the state started to intervene. The pancadão then morphed into a third version. This version, the topic of this chapter, is what I call the "mobile pancadão." Rather than depending on a mounted stage, the party organizers started to revolve around the car audio. This format allowed partygoers to literally move the event to unattended streets whenever the police tried to dismantle them. Moreno describes a mobile pancadão:

> [. . .] The cars arrive, always with a male youth driving (alone or with another male friend) bringing girls to the party who often hang outside the car windows to get attention. Some [of the girls] [. . .] show up with almost the whole body outside the car, dancing to the sound of *funk*. When the car

stops, the boys open the trunk [...]. The sound is turned to maximum and the girls begin a new choreography around the car. This repeats with all cars that arrive and park behind or next to each other, each with the sound on [...]. Sometimes the [male] drivers take part in the choreographies, but the movements are different, signaling sexual intercourse. [...] The girls always stay close to the car. (Moreno 2011)

Coordinating the mobile pancadões has become possible with the widespread access to mobile Internet access in Brazil.[11] Whenever partygoers find out the police might disrupt the pancadão, they use Twitter and WhatsApp[12] to move the gathering to another location. Because of these dynamics, party organizers continuously monitor potential spaces for future pancadões in their neighborhoods—particularly narrow streets and obscure plazas where the police usually don't go.

There are a few links between the mobile pancadão and funk ostentação worth mentioning. First, the party format enables the car owner's *sonic* ostentation in addition to the visual and musical ostentation embedded in the song. At the mobile pancadões, the music that talks about ostentation comes from an ostentatious device (the audio equipment), which is integrated into another ostentatious device (the vehicle)—a triple ostentatious move. Moreover, the funk ostentação songs, which describe cars as vehicles of ostentation and status, are coming from cars, imbuing the car owners with the MC persona (itself the wealthier version of the poor suburban youth). Framing space through its acoustic domain, the car establishes the party environment and mediates social interactions and status through the concentric propagation of powerful beats.

Kondzilla explains that funk ostentação drew on Houston hip-hop to build an audiovisual narrative of sonic dominance, street authenticity, and hedonic ostentation. The importance of automobility in Houston hip-hop relates to hustling, with rappers often incorporating the pimp persona and lifestyle (people who live on and for the streets) in music videos. In this narrative, as the pimp's low-rider moves slowly and assertively from one transaction to the other, the booming

---

[11] In the past most of the access was centralized at "LAN houses," a venue that offered computer usage with Internet for a per-minute fee. As faster cable or DSL connection spread out from the center and reached the suburbs, those with access started to redistribute connectivity informally. Owners of illegal Internet redistribution points pay for fast-speed legal Internet connection and install a router to send the Internet signal to a Multichannel Multipoint Distribution Service (MMDS) that encodes the signal into radio waves. The microwave antenna connected to the MMDS device sends the signal to a relay antenna installed at a higher point, which allows the signal to reach households located up to 11 yards from it. The final users pay around $100 for a MMDS (to decode the signal) and a modem, plus a $25 monthly payment (Domingos and Carpanez 2009).

[12] *WhatsApp* is an instant messaging application for smartphones. In 2012, it allowed users to join up to fifty group chats, each with a maximum of fifty users.

music becomes a device to claim dominance over his turf and over the women who work for him. Sub-bass music (in the range of 25 and 50 Hz) is thus an acoustic red carpet, allowing the pimp-driver to be noticed well before his visual entrance.

In São Paulo, a city where car culture has always been prevalent, the new centrality of cars within the funk ostentação narrative promoted further shifts in the music. The long sub-bass notes popularized by Houston hip-hop were incorporated in funk ostentação.[13] According to funk carioca DJ Tecyo Queiroz, "Funk ostentação is becoming less about dancing and more about the car. With funk, ostentação they started using the prolonged bass drum to make everything vibrate" (April 2013, personal communication). A large mobile pancadão includes several cars booming different funk ostentação songs a few feet from each other. The more vehicles, people, and loud grooves present, the more the public space is stabilized and translated into a party environment. Loudness is particularly important here because the parties take place outdoors, with little reverberation.

Most pancadão vehicles have two 15-inch woofers (or subwoofers), one to four mid-range speakers, and two to four tweeters for higher frequencies. Santa Efigenia Street in downtown São Paulo is the mecca for car audio equipment, with dozens of busy stores selling speakers, amplifiers, receivers, and so on. Installation of car audio equipment usually takes place in small garages, which can be found not far from Santa Efigência Street. In these stores, money is translated into car audio power, which is measured into RMS values in watts.[14] If the New Acoustics described in Chapter 2 played a role in the engineering of audio transmission and noise measurement, it also helped to consolidate more efficient car audio machinery. The list of black boxes necessary to translate electric energy into sound waves also includes an amplifier (to increase the energy of the input signal), equalizers (to alter the intensity of specific frequencies), crossovers (to separate optimal frequency bands for each speaker), capacitors (to deliver extra energy in case there is a signal peak, as subwoofers are energy-hungry), and damping material (trunk enclosure that allows woofers and subwoofers to generate lengthy low-frequency waves more effectively). In 2012, a car owner could transform his vehicle into a resonant device fit for pancadões for around US$1,200 (see Figure 6.1).

---

[13] This is an important sonic shift in relation to funk carioca, which usually has a "dry" bass drum beat in its mixes—the association with "big thump" or punch refers to this sonic quality. This short bass drum, which resembles the sound of the samba bass drum, was considered particularly practical for dancing indoors, where most funk parties in Rio take place.

[14] RMS (root mean square) is a method for measuring the amplifier's average output power. It denotes how much electrical energy is required to send to the speakers without distorting or overheating them. The RMS equation "takes into account the power sent from the amplifier, the impedance of the current and the inductance of the magnetic field of speaker current and the current itself." http://www.ehow.com/facts_4899035_what-does-rms-stand-speakers.html

The "Rowdy" Teenagers | 179

FIGURE 6.1. A typical pancadão resonant machine, with two subwoofers, one mid-range speaker, and four tweeters.
Photo by the author, March 2012.

The mobile pancadão is part of a global car industry that targets male car owners interested in "dominating" the street as they move around. For instance, in 2004 Pioneer Electronics launched a series of in-car sound equipment ads with the slogans "Defy," "Disrupt," "Disturb," and "Ignite." The US$3 million controversial campaign, targeted at sixteen-to-twenty-four-year-old "Gen Y" male consumers "with a passion for cars and entertainment,"[15] showed the stories of male youths who had invested thousands of dollars in Pioneer audio equipment. The video ads show young males making statements such as "You are going to hear what I'm playing regardless of whether you like it or not" and "I was like a mini earthquake driving through."

Brendon LaBelle claims that the tradition of turning entire cars into mobile sound systems originated mainly within Mexican American and African American youth groups for whom "the street is [. . .] a kind of stage for the production of

---

[15] "Pioneer Offers Attitude and Information to Gen Y Males with an Edgy Print and Web-Oriented Advertising Campaign," https://www.pioneerelectronics.com/PUSA/Press-Room/Car-Audio-Video/Pioneer+Offers+Attitude+And+Information+To+Gen+Y+Males+With+An+Edgy+Print+And+Web-Oriented+Advertising+Campaign.

beats that extends the skin—of both the drum and the body—to the resonating mold of the street: a kind of beating back to the violence streets have come to force onto Mexican American and African American youths" (LaBelle 2008, 197). Paul Gilroy argues that, as spatially segregated youths consume automobility, the car emerges as a "giant armored bed on wheels that can shout to the driver's dwindling claims upon the world into dead public space at ever-increasing volume" (Gilroy 2001, 97). Ironically, the very object that through most of the twentieth century was used by the urban elites "as a means of physically separating [themselves] from spatial configuration like higher urban density, public space, or from the city altogether" (Henderson 2006, 294) became the device the segregated youth deploy to create leisure zones capable of unsettling middle- and upper-class residents.

Renato Barreiros, the young subprefect who organized the first Funk Festival in Cidade Tiradentes, explains that funk ostentação exploded in the late 2000s because it became what he calls the "chronicle of the new periphery":

> "Bonde da Juju" describes the clothes and apparel that local kids liked to wear. After narrating what they already had, MCs started to go after other objects of desire, trends, things they would like to have. And why didn't this happen in the 1990s? In the 1990s inflation and unemployment were high and purchasing power was low. There was no perspective. The young generation is living in a moment of well-being, starting to consume things they never imagined they would. (Personal communication, October 2012)

A 2013 report published by the federal government suggested that between 2002 and 2012, 35 million Brazilians left critical poverty levels to become part of the "class C," making between US\$720 and US\$1,875[16] (Grosner et al. 2013). Barreiros argues that in the 2000s, the emergence of a wealthier class C caused a sort of "identity crisis" among middle-class Brazilians, as members of class C started to have access to plane tickets, cable TV, Ed Hardy clothing, and smartphones. It seemed the middle class had lost the exclusive privilege of spatial mobility and consumption as status-claim. Could we argue that peripheral youth embrace the pancadão as a way to affirm their citizenship as consumers? Is the pancadão a version of citizenship that, as Nestor Garcia Canclini puts it, is expressed "more often than not through private consumption of commodities and media offering than through the abstract rules of democracy or through participation in discredited

---

[16] According to the Brazil for Foreigners website, the monthly household income for so-called class C residents is between R\$2,040 and R\$5,100, equivalent to approximately US\$720 to US\$1,875 at the time of writing.

political organizations" (Canclini 2001, 5)? Like the pixadores and the motoboys examined by Caldeira, the pancadão participants are challenging the limits of democratization and engaging in ambiguous insurgent citizenship claims.

Disarticulating the Pancadão

In 1990s Rio de Janeiro, as funk carioca entered the group of Latin American music genres coming from the low-income margins and deemed "noisy,"[17] a discourse of moral panic emerged in the press. Residents worried that the black male teenagers from the *favelas* were "invading" the luxurious beaches in southern Rio to rob tourists. The authorities and the press stated that the young robbers were drug users and *funkeiros* (fans of funk carioca). In 2000, under the argument that fights between partygoers and the presence of "pornographic," anti-police songs in funk parties were "corrupting" youth behavior, the Rio de Janeiro legislature passed a law[18] that required party organizers to have police consent in written form beforehand, and have police officers present during the entire event (who could stop the party at any sign of aggressive or lewd behavior). The law also prohibited sound crews from playing songs that praised criminal activity or drug factions.[19] This

---

[17] I am referring, for instance, to the 1910s, when the Rio de Janeiro government prohibited the performance of Afro-Brazilian music due to their religious and "primitivistic" connotations, at a moment when the Rio de Janeiro authorities were trying to emulate European model of civility (see Sandroni 2001). Like funk carioca, reggaeton and the *perreo* dance associated with it have been condemned in many parts of Hispanic Latin America. Framed as the sign of an irrepressible, vulgar, and disrespectful youth generation, reggaeton has been attacked by a wide range of groups concerned with the degradation of "family values," "good taste," and "ethical standards." In 2002, senator Velda González, chairperson of the Special Commission for the Study of Violent and Sexual Content in Puerto Rican Radio and Television, organized public hearings to address "indecency" and "pornography" of reggaeton music videos. In 2012, the Cuban government banned reggaeton for its "aggressive and sexually obscene lyrics" (quoted in Wagner 2012). According to Orlando Vistel Columbié, head of the Cuban Music Institute, "We are in the process of purging music catalogs with the aim of eradicating practices that, in their content, stray from the legitimacy of Cuban popular culture" (quoted in Tremlett 2012). The crackdown included professional disqualification, and sanctions against government officials who encouraged reggaeton performances.

[18] Rio de Janeiro State Law 3410/2000.

[19] Party organizers, community leaders, and sound teams criticized the law as discriminatory. As George Yúdice (2003) has observed, during the 2000s funk carioca in Rio de Janeiro went through a process of decriminalization. Its articulators started to claim that its expression (both as music and party) was a cultural right. These groups framed funk carioca artists and its fans not as potential criminals, but as victims of cultural prejudice and police abuse. This approach pressured the local government to provide leisure spaces for funk parties and revise its own taste preconceptions. In 2003, three years after Law 3410 was created, the state of Rio de Janeiro passed a bill for promoting funk parties as a "cultural activity of popular character." In 2008, when funk ostentação exploded in Cidade Tiradentes, Rio de Janeiro lawmakers decided to terminate Law 3410/2000 because of its discriminatory overtones.

move contributed to the further migration of the parties to *favelas*, where state surveillance has been difficult, to say the least.

Similarly, as funk ostentação spread out across São Paulo from Cidade Tiradentes using online and offline networks, the press alerted the public to the criminal ramifications of the pancadão. The stories emphasized crime and drug consumption as signs of suburban youth lifestyle directly linked to the pancadões. In 2008, *O Estado de São Paulo* published a story about minors using drugs at a pancadão in front of a public school. A 2011 story by the same newspaper claimed that minors freely consumed alcohol at pancadões. The story quotes one drug dealer attending the party stating that the profit with drug commerce at pancadões was twice what he made on a regular day. One fourteen-year-old female smoking marijuana disclosed she expected to have sex at the party. The press also described the harmful effects of loud grooves. The following excerpt illustrates a recurrent narrative:

> The party is chaotic. Cars with powerful sound equipment park on the streets where the event is happening [. . .]. Each car plays a different funk [carioca] song at the maximum volume. Last Saturday six cars animated the party [. . .] until 8:00 or 9:00 a.m. on Sunday. The residents live under stress. They can't sleep on Saturday nights. Besides the hellish noise, they can't get in or out of their own garages because the crowds occupy the driveways.[20]

The CONSEG Meetings

In Chapter 4, I described the Community Security Council (*Conselho Comunitário de Segurança*, or CONSEG) and its importance in connecting residents with law-enforcement agencies. I argued that the assertive presence of the military police was directly related to how Paulistanos experience their acoustic environment. In 2012, I visited the archives at São Paulo State Public Security Secretary[21] and read the minutes from the CONSEG meetings between 2010 and 2012 (or the latest available month).[22] In reading the minutes, I was surprised by the increase in noise

---

[20] "Tráfico mantém baile funk em Heliópolis," *O Estado de São Paulo* (November 30, 2011), C5.
[21] The Public Security Secretary manages the military and civil police. The head of the secretary is nominated by the state governor.
[22] These data present some problems of reliability. A secretary who is part of the local community is responsible for writing the minutes. There is little consistency in the amount or type of information included in the minutes. Some include dozens complaints made by the residents, whereas others are extremely short and only include what the authorities said. For instance, although some meetings in Cidade Tiradentes in 2010 and 2011 had more than one hundred attendees, the minutes do not include specific complaints.

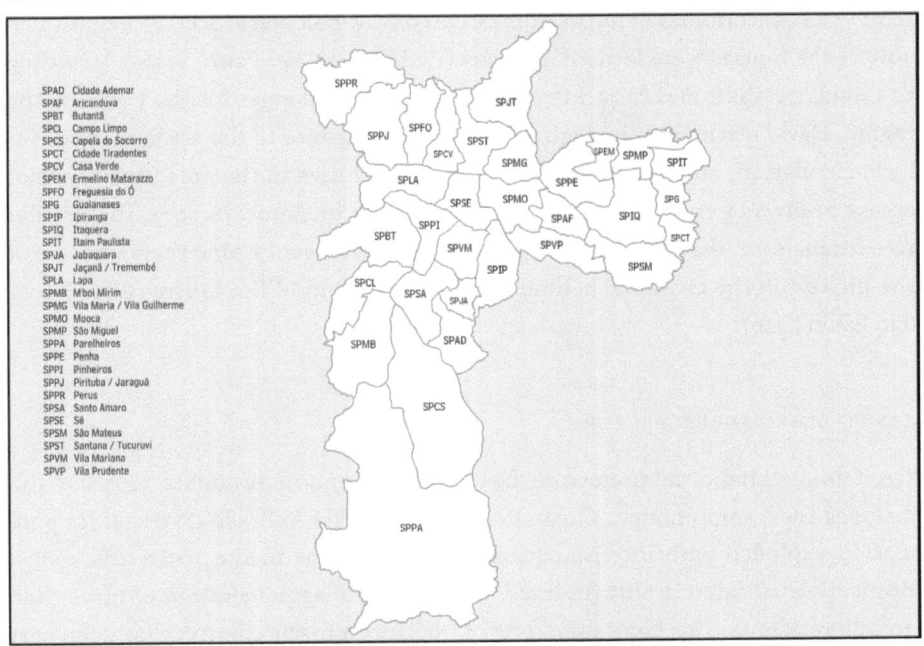

FIGURE 6.2. Map of the city of São Paulo showing the subprefectures. Unlike the previous chapters in this book, which take place in the Sé subprefecture, here we consider the Ipiranga, Jabaquara, Campo Limpo, and M'Boi Mirim subprefectures.

complaints related to the pancadão. The archival research focuses on the CONSEG meetings in locations I was able to attend during fieldwork; it encompasses four subprefectures: Jabaquara, Campo Limpo, Ipiranga, and M'Boi Mirim (see Figure 6.2).

JABAQUARA SUBPREFECTURE

Jabaquara is a one-district subprefecture. It covers an area of 7.5 square miles and has 223,000 residents, with a population density of 15.8 inhabitants per square kilometer (São Paulo City Hall Website 2017). Police officer Captain David, who is in charge of a police station in Jabaquara, says that the "fluctuating population" (residents plus those who circulate through Jabaquara during work hours) inflates this figure to 550,000 people. Jabaquara territory includes two metro stations, a bus station, the Congonhas airport, and a few administrative buildings. It can be divided into three major socioeconomic zones: middle-class condominiums and houses; the central business area, which has access to public transport; and the slums and recently re-urbanized areas. David says that there are thirty-five slums in Jabaquara. In the last few years, the state and municipal governments have removed some families living in slums located in at-risk areas. Still, in 2015 almost 20% of the houses were located in slums (Rede Nossa São Paulo 2016). According to David, car theft and carjacking are the biggest challenges for the police in the region. David attributed the high number of auto crimes to the area's massive vehicle circulation, an increase in the number of vehicles in the area (due to easier access to lines of credit), and the area's proximity to slums. In 2011, the murder rate for male youths between the ages of fifteen and twenty-nine years was one of the highest in the city, only behind Campo Limpo and M'Boi Mirim (Rede Nossa São Paulo 2016).

CAMPO LIMPO SUBPREFECTURE

The Campo Limpo subprefecture has roughly 607,000 residents (2010)[23] and includes the Campo Limpo, Capão Redondo, and Vila Andrade districts. Its population exploded with intense squatting in idle farms in the 1970s and 1980s. Paraisópolis, located in Vila Andrade, is an informal agglomeration of more than 40,000 residents. This large slum (one of the largest in the country) has been part of class tensions due to its proximity to the upscale Morumbi district. In the 1990s, the Campo Limpo district has been labeled, together with Capão Redondo and Parque Santo Antônio, São Paulo's "death triangle," due to the high homicide rate.[24] In the last decades, the proximity to the businesses on the Pinheiros Expressway and Berrini Avenue stimulated the construction of middle-class residences in the region. Many of the upper-class residents who moved to the region live in gated communities and condominiums.

IPIRANGA SUBPREFECTURE

The Ipiranga subprefecture includes the Ipiranga, Sacomã, and Cursino districts. Heliópolis, a large community within the Sacomã district, became known in the 1990s as one of the largest slum complexes in Brazil. In the 1970s, immigrants from the Northeast moved to Heliópolis after being evicted from "areas of interest" by the municipal administration. At that point in time, the region did not offer infrastructure or public services, which meant that these newcomers had to

---

[23] "População Recenseada, Taxas de Crescimento Populacional e Densidade Demográfica: Município de São Paulo, Subprefeituras e Distritos Municipais," https://www.dropbox.com/s/pxnm17u7i1phrs3/populacao%20recenseada%20sp.pdf?dl=0.

[24] "'Triângulo da morte' concentra assassinatos em SP; veja mapa," *Folha de São Paulo* (January 26, 2012), http://www1.folha.uol.com.br/cotidiano/2012/01/1039884-triangulo-da-morte-concentra-assassinatos-em-sp-veja-mapa.shtml.

build their own houses from scratch. Today, a large part of Heliópolis residents are immigrants from the Northeast of Brazil (UNAS n.d.). More recently, Heliópolis has been going through a process of state-sponsored urbanization, what the authorities call "re-urbanization." In the last decade, both the Municipal Housing Agency and the Housing and Urban Development Company of the state of São Paulo have sponsored housing projects. Whereas "re-urbanization" has started to increase land value and push poor families further to the outskirts of the city, most households are still defined as slums. Like the other districts considered here, the area has a large youth population: in the late 2000s, roughly 52% of those living in Heliópolis were twenty-five years old or younger (Brancatelli 2009).

M'BOI MIRIM SUBPREFECTURE

The M'Boi Mirim subprefecture includes the Jardim Ângela and Jardim São Luís districts, comprising 563,000 residents (2010).[25] As Heliópolis, M'Boi Mirim was occupied by immigrants looking for jobs, particularly in the 1970s during the city's peak of industrial activity. According to Beto Mendes, the subprefect of M'Boi Mirim between 2009 and 2012,[26] "Today we are in a process of re-urbanization, bringing water, sewer, and basic sanitation to these communities, removing families from areas at geological or flooding risks. Up until recently, there were 270 slums in the region. Today, we have a little more than 100" (personal communication, September 2012).

In 1996, Jardim Ângela was labeled one of the most dangerous districts in the world, with 130 deaths per 100,000 inhabitants. Although the number of homicides has decreased considerably in São Paulo, between 2003 and 2011 M'Boi Mirim was in the top-five list of subprefectures with the highest number of homicides of men between fifteen and twenty-nine years old. Mendes argues that pancadões became popular because organized crime in the region takes advantage of youth idleness:

> The youth understand that there are not many alternatives for entertainment here. Ill-intentioned people take advantage of this situation to organize events outdoors, on the streets, disturbing the peace of families—especially those families that need to wake up early on the next day. Imagine living next to a plaza, with a pancadão (with that uproar) going on in front of your house

---

[25] "População Recenseada, Taxas de Crescimento Populacional e Densidade Demográfica," https://www.dropbox.com/s/pxnm17u7i1phrs3/populacao%20recenseada%20sp.pdf?dl=0.

[26] Mendes would be removed from the subprefecture in November 2012, a few months after I interviewed him. The São Paulo State Court found him guilty of not transferring the social security contribution of municipal employees between 1997 and 2004 during his term as mayor of Paranapanema.

and you having to wake up early to go to work. (Personal communication, September 2012)

Table 6.1 draws on the minutes and compares the frequency of four problems regularly discussed at the meetings: theft, the pancadão, other sound-related problems (bars, school, commerce, etc.), and dirt. Theft, which includes mostly car and home theft, was the most common complaint. Dirt was also a regular concern and is included here because it was often framed together with community noise as a sign of disorder and a critical indication of the relationship between the community and the state.

In 2012 and 2013, I attended several CONSEG meetings in these four subprefectures to observe the reactions of residents living close to places where pancadões occurred regularly. People from different socioeconomic backgrounds often attend the CONSEG meetings. All attendants live or work in the region and are interested in resolving specific problems in their neighborhood. Some people attend the meetings regularly as they continue to experience problems that need to be solved. Others (a minority) are always present, even without specific complaints to make. Some stop attending meetings when their problem is solved—an attitude that is publicly reproached by the regulars during the meetings. Below I include some exchanges that took place during the meetings I have attended. They should give the reader a sense of how residents attempted to mobilize the state through narratives involving civility, noise, and youth transgression. The format I present these exchanges is the same as in Chapter 2. I start with quotes from the meetings related to a specific issue to give the reader a better sense of how the actors interacted, and then move to unpack some elements of the interactions.

PANCADÃO AND THE RACE-CLASS-SPACE CONTINUUM

At a CONSEG meeting in Heliópolis:

> Resident 1: *On Saturday, I called 190 [police emergency number] and told the police there was a guy blasting music from his car with his doors opened. They left me waiting on the phone for fifteen minutes. Then a policeman picked up the phone and asked me: "What's the matter? [...] "Oh ... OK, a police car will go shortly." Nobody showed up. The police are prejudiced because it's a slum. They don't go there!*

A participant at the CONSEG meeting in Campo Limpo asks the military police to explain an action undertaken shortly before the meeting to disperse a pancadão:

TABLE 6.1.

Comparison of the frequency of problems discussed at CONSEG meetings in five districts since 2010. Based on analysis of minutes archived at the São Paulo State Public Security Secretary.

Jabaquara Subprefecture—CONSEG meetings

Problem[a]	2010 (Feb.–Nov.)	2011 (Feb.–Nov.)	2012 (Feb.–May)
Pancadão	10%	27%	31%
Other sounds	13%	9%	14%
Theft	50%	49%	48%
Dirt	27%	15%	7%

Campo Limpo Subprefecture – Campo Limpo CONSEG meetings

Problem	2010	2011	2012 (Jan.–April)
Pancadão	11%	23%	20%
Other sounds	7%	7%	16%
Theft	67%	51%	45%
Dirt	15%	19%	19%

Campo Limpo Subprefecture – Capão Redondo CONSEG meetings

Problem	2010	2011 (Jan.–Dec.)	2012
Pancadão	33%	38%	–
Other sounds	11%	5%	–
Theft	39%	38%	–
Dirt	17%	19%	–

M'Boi Mirim Subprefecture—Jardim São Luiz CONSEG meetings

Problem	2010 (March–Dec.)	2011 (Feb.–Dec.)	2012
Pancadão	19%	32%	–
Other sounds	4%	8%	–
Theft	35%	46%	–
Dirt	42%	14%	–

Ipiranga Subprefecture—Heliópolis CONSEG meetings

Problem	2010 (Jan.–Nov.)	2011 (Jan.–May)	2012
Pancadão	31%	33%	–
Other sounds	19%	30%	–
Theft	44%	20%	–
Dirt	6%	17%	–

[a] The percentage reflects the proportional relations between the four categories analyzed.

Sergeant (Military Police): *This action was actually not on our schedule. But in the last two months, we had been receiving several complaints saying that we don't go there* [i.e., to the slum]. *So, we went there that night with nine police cars. When we got there, there were at least 1,000 people at the pancadão.... We started to charge. People who go to pancadões don't like the military police. They threw stones and empty bottles at us. So, we took action. After five minutes, the place was empty... We put an end to the pancadão* [several attendees applaud].

At a CONSEG meeting in Jabaquara:

Resident 2: *I have a problem with these cars with loud sound systems. I've been complaining for more than a year. We call the police, we call whomever. . . . We can't watch TV, sleep, or study.*

Council President: *Where is your apartment? Next to the slums?*

Resident 2: *Yes, next to the slums. We have nothing against them, but they have to respect us.*

Those living in impoverished areas often feel discriminated against by other CONSEG participants and by the police. Whereas some participants suggest that the lack of infrastructure in the slums encourages disorderly and illicit activities such as pancadões, those who live in the slums argue that it is precisely the absence of the authorities that makes these places a "no man's land" that attracts pancadões. The relationship between space and class was expressed more regularly than that between space and race, which suggest less an absence of racism than the subtle mechanism of reading space as embedded in implicit race and class distinctions.

Some police officers openly explained why they were not willing to deal with pancadões in the slums. At a meeting in Campo Limpo, the captain said that the police would only help the community with disturbances of the peace if they came to the CONSEG meetings and publicly asked for help. His argument was that without the support of the local community, police operations to crack down on *pancadões* would not be "well-received": "We are going to get bottles thrown at us, we are going to get shot. . . And the next day the community will appear on TV to call us animals." Here community participation is taken as a requirement for the right to public security: for the police to take action in a given area, the captain requested the establishment of mutual trust. In several occasions, police officers suggested that preserving the image of the police and the security of police officers and their equipment were important factors when deciding how, where, and when to act.

PANCADÃO AND FAMILY VALUES

At a CONSEG meeting in Campo Limpo:

> Resident 3: *I was at the pancadão. Everybody was dancing, having fun, and listening to music. Then suddenly, the police arrived. They invaded, throwing bombs, shooting rubber bullets, beating... Little kids got hurt; a pregnant woman was beaten...*
>
> [General uproar, many people outraged. Someone screams, "And what was a pregnant woman doing at a pancadão?"]
>
> Resident 4 (from the same group as Resident 3): *Whether we like it or not, we have to listen to what others have to say. I don't like pancadões. . . . everybody knows that. I hate music that doesn't add anything—music with no quality; I tell this to my sons. But I think what happened here was a total lack of respect. We need to listen to everybody. The girl was not able to finish what she had to say because she was booed at a CONSEG meeting. That is a shame for us. The youth have a right to have fun, a right to have culture, to practice sports. What has the subprefecture done for leisure in our region? Nothing. The subprefecture could build hangars to promote parties with security. Then we wouldn't have to sweep cocaine glass holders from our front doors.*

At a CONSEG meeting in Jabaquara:

> Resident 5: *We were young, and we didn't have a place to go. "Oh, those poor souls, they cause trouble because they don't have a place to hang out" ... that's an excuse! And did we have a place to go? No, but we showed more respect.*
>
> Another resident in the same conversation: *The hospital is receiving pregnant girls... They're all coming from these funk parties... Because every corner and dark alley is a place to have sex...*

At a CONSEG meeting in Heliópolis:

> Captain (Military Police): *Funk parties have dissipated throughout São Paulo. We already lost a generation. Unfortunately, I have to say we have lost a generation. A generation involved with funk parties, drugs... Nobody can recover.*

The vast majority of the participants at the CONSEG meetings I attended were forty years or older. There was a general perception among them that a "generational shift" had taken place. They argued in various ways that youths had lost respect for the established authorities such as parents, police, and teachers. The youth was a "disruption" in the community and the police were seen as the only group with enough authority to put them in their "proper place." The quote above, from a young female partygoer complaining about the use of force by the police at a pancadão, was the only case I witnessed at a CONSEG meeting of someone talking openly from the other side of the controversy. The indignant reaction of other participants when she tried to make a complaint is telling. For most people at the meeting, it was not inconceivable that a policeman might have beaten a pregnant woman, but that this woman would be at a pancadão, the epicenter of disorder and crime. For many CONSEG participants, poor youths are potentially the personification of disorder not only because of their habits of loitering but also because they "take advantage" of the legal system. These participants—including CONSEG board members such as police officers and commissioners—blame the judiciary and legislature for "overprotecting" the adolescents.

Caldeira defines "talk of crime" as everyday conversations about fear and crime. Highly contagious, this type of talk "feeds a circle in which fear is both dealt with and reproduced, and violence is both counteracted and magnified" (Caldeira 2000, 19). CONSEG meetings are fertile spaces for turning the talk of crime into demands for local change. Because complainers often magnify the criminal component of pancadões at these meetings, the authorities are encouraged to treat the parties as a public security problem. As we saw in the previous chapters, zero tolerance was popular during Rudolph Giuliani's administration in New York City (1993–2001). William Bratton, appointed by Giuliani as the New York police commissioner, drew on a specific reading of broken windows theory to establish zero tolerance measures. Bratton's zero-tolerance policing drew on Wilson and Kelling's argument that "disorder and crime are usually inextricably linked, in a kind of developmental sequence" (1982). Under Bratton, the NYPD became invested in removing homeless people from the subways, "squeegee pests" from stoplights, and panhandlers, drunks, and "noisy" teenagers from the sidewalks (Bratton 1998, 33). For Bratton, zero tolerance policing was crucial to the revitalization of the city: "New Yorkers are reporting that they are feeling safer. Residential and commercial real estate markets are booming. The economy has stabilized. Tourism is skyrocketing. New York City is slowly revitalizing itself" (Bratton 1998, 41).

Zero tolerance was particularly important for Colonel Álvaro Camilo, Chief of the São Paulo State Military Police between 2009 and 2012. Camilo argues that the pancadão is part of a "culture of disorder" and therefore a public security issue. As Camilo explains, "To allow the youth to commit some types of offenses is an incentive to impunity. Crime must be fought from side to side" (quoted in Rabelo

2015). Camilo assigned the military police battalions across the state the task of disarticulating the pancadões.

In the Introduction, I argued that the order presented in this book (legislative, executive, and judiciary branches) should not be read as an indication of how governments operate. So far, this order has made the narrative smoother, as we have examined controversies that draw on each other in a logical progression (technical standards, noise ordinances, law enforcement, and litigation). The next section gives us an opportunity to see how noise can reorganize the state. To do that, I move backward through the book and examine how the pancadão has rewired the actor-networks discussed in the previous chapters.

REWIRING LEGAL CHANNELS

In 2011, Lieutenant-Colonel Deufrânio Barbosa de Carvalho became head of the 37th Military Police Battalion (BPM), which oversees parts of Campo Limpo and M'Boi Mirim—an area that comprises roughly 900,000 people and 111 slums.[27] A few months after taking over the 37th BPM, Carvalho received a letter from the state prosecutor. In the letter, the state prosecutor stated he had just attended a CONSEG meeting and, from what he had heard, it seemed that the police had lost control over the pancadão, an event with "loud sounds and adolescents consuming drugs and alcohol."[28]

The prosecutor explained that the existing legislation gave the police the proper legal channels to tame the pancadões. One alternative was Article 228 from the 1997 Brazilian Traffic Code, which prohibits "sound equipment in the vehicle at a volume or frequency not authorized by the CONTRAN." The CONTRAN (National Traffic Council) Resolution 204/2006, which draws on the two CONAMA 1990 resolutions (discussed in Chapter 2), establishes as a grave infraction[29] any sound equipment that exceeds 80 dB, measured approximately 7.5 yards from the car. As a M'Boi Mirim subprefecture official admitted, "Any car audio that makes a sound over 80 dB is considered a mischaracterization of the vehicle. That, of course, is a matter of interpretation, or we would have confiscated half the cars circulating in São Paulo!" Although it seems the official was mistakenly mixing different articles

---

[27] "Histórico," *370 Batalhão da Polícia Militar do Estado de São Paulo*, http://www.policiamilitar.sp.gov.br/unidades/37bpmm/historico.html.
[28] This quote is from the state prosecuter's letter, a copy of which I received from Carvalho.
[29] The Brazilian traffic code operates as a point system. If a driver accumulates more than twenty points, the driving license is suspended. An infraction ranges from three to seven points. A *grave infraction* adds five points plus a US$55 fine.

from the Brazilian Traffic Code,[30] her argument holds up: the executive would have to fine many more car owners if it enforced the law in non-pancadão contexts. Since sound measurement was often the only legal justification that the police had to punish the car owner, and since the military police were not equipped with sound level meters, the PSIU would be crucial in legitimizing the police action. However, from the authorities' point of view, this type of offense had two major drawbacks: it did not determine the confiscation of the vehicle, and the fine was too low.

After discussing other legal possibilities, Lieutenant-Colonel Carvalho and the state prosecutor agreed it would be more efficient to take another legal channel by interpreting the sound from pancadão cars as an environmental crime. As we learned from the previous chapter, although this is the most punishable infraction for environmental noise, it is also a risky one because the judiciary is reluctant to tie noise to the Environmental Crime Law. For the executive, using this channel gives them the ability to apprehend the pancadão vehicle right away for forensics. With the car confiscated it would be possible to maximize the financial damage done to the car owner. As the following statement from a civil police commissioner during a CONSEG meeting in the Jabaquara district suggests, the strategy of causing financial pain to noisy car owners was common and described openly:

> When I was working in another district, we apprehended a few pancadão cars. But the judiciary ordered us to return the cars with a petition. I went to the judiciary and explained to the judge that the situation was serious. Yes, if everything is okay with the car, we need to return it to the owner. But here's the tweak: we did the forensics in the car with *no* hurry. Sometimes, we would take ten, twelve days to do it, keeping the car in our garage. That's what the police need to do . . . wear the car owners out financially.

REWIRING ADMINISTRATIVE FLOWS

Having rewired (as much as possible) the legal channels to make sure the police could act against the pancadões, Carvalho appointed a group of police officers to collect all the necessary data about the parties. The officers compiled the dossier: "Assessment of Improvised 'Funk' Parties in São Paulo and Metropolitan Region,"[31] which mapped more than three hundred pancadão points across the

---

[30] Article 228 regulates loudness; changes in the characteristics of vehicles are regulated by Articles 96 and 106.
[31] Lieutenant-Colonel de Carvalho kindly gave me a copy of the "funk carioca dossier" created by the 37th BPM.

city. Since each public institution had legal limitations on tackling the pancadão, Carvalho proposed, in collaboration with the M'Boi Mirim subprefecture, an integrated task force. The task force would include food supply inspectors, the PSIU, the military police, the civil police, the Metropolitan Civil Guard, the Guardianship Council, and the traffic agency. In that way, the military police and the subprefecture hoped to avoid any potential clogging of the administrative flows to dismantle the pancadão. The M'Boi Mirim subprefecture shared with the other institutions the report "Battle Against Pancadões,"[32] a document that detailed how and when each participant would contribute in the operation.

The subprefecture would provide the municipal agents who would inspect bars (e.g., hygiene regulation) and confiscate the street vendors' equipment and products. Metropolitan civil guards would be responsible for protecting the city's estate. They would also protect the municipal inspectors and, if necessary, preserve the (potential) crime scene for further forensic investigation. Guardianship Council agents would mediate between the juvenile offenders and the state by dealing on-the-ground with the underaged partygoers. The 1990 Child and Adolescent Statute (*Estatudo da Criança e do Adolescente*, or ECA) establishes that, due to their condition as "people under development," children and adolescents cannot be arrested, only detained. During the anti-pancadão operation, minors would be taken to a police station, where a Guardianship Council agent would be present. The military police would decide where the operations would take place, relying on 190 calls, CONSEG meetings, data from other institutions, and undercover police officers infiltrated at the parties. The military police would also disperse the crowd and detain anyone potentially or blatantly involved in illegal activities, verify the car owners' documents, and apprehend the vehicle. The PSIU agents would measure sounds from vehicles.

Carvalho explains that the strategy for disarticulating the pancadão involves the strategic use of nonlethal weapons. The police deploy the Riot Control Group to deal with "mob" dispersal and to protect police vehicles and equipment from the rocks and glass bottles thrown by the more defiant partygoers. As the Riot Control Battalion marches on with riot shields and nightsticks, a group of three or four officers uses smoke and stun grenades to disperse the party. With the ground "regained," as the police officers check documents, the PSIU and subprefecture agents inspect venues and vehicles.

The first operation in M'Boi Mirim involved forty civil guards in thirteen cars and forty-one military police officers (including members of the Special Unit and Riot Control Battalion) in thirteen cars, four motorcycles, and two mobile stations.

---

[32] Mendes's secretary kindly gave me a copy of this document.

A total of 120 people from the mentioned institutions participated in this operation. The task force confiscated four cars, fined four bars, and detained forty-two minors. On the next day, *Folha de São Paulo* issued a half-page story describing the operation and the success of its administrative rewiring (see Figure 6.3).

In 2012, I accompanied several anti-pancadão operations with the PSIU. Unlike the regular PSIU inspections described in Chapter 4, the PSIU would not know beforehand where the operations would take place or how long they would take, as the military police coordinated the task force based on information received minute-by-minute. Below I narrate an operation that took place on March 25,

FIGURE 6.3. The half-page story published in *Folha de São Paulo* about the first anti-pancadão operation in early 2012, with a photo of detained teenagers. The main headlines read: "Operation closes funk party and takes 42 minors to the police department." Below: "The city suspends a pancadão on the south side with the support of the military police, municipal agents, and guardianship counselors."

Credit: *Folha de São Paulo* (January 31, 2012), C5.

2012, which involved the M'Boi Mirim subprefecture and Carvalho's 37th Military Police Battalion.

At a closed meeting with all the public officials who would participate in the anti-pancadão operation that night, we are in a small room facing a whiteboard where the lieutenant who is in charge of the whole operation has drawn a schematic map of the location where the pancadão is taking place:

> MILITARY POLICE LIEUTENANT, ADDRESSING THE PSIU CREW: *To ensure your safety, you're going to stay here at the military police base with some police officers. As soon as we take ground, and already have everything under control, I'll authorize them to come with you. That's also because if we go out on a very large convoy, the wireless communication will arrive there, and then we'll lose the operation. We will leave in three convoys. One team is going to go down this street [he points to the whiteboard]. Team 2 will come here, and the third will come here, from below. We are going to leave this street for them to leave because we have to leave an escape route. If we don't leave a route they are going to charge at us, and that's not our intention. Our intention is to catch the vehicles, which are the cause of this whole situation.*
>
> SUBPREFECT: *We seized nine vehicles in the last operation. So far only three are trying to get their vehicles back. This causes them big financial damage; I believe the other six will not even try to get their cars released because it's not worth it, with all the fines. For the subprefecture, it is important to take hold of the largest number of street vendors too. . . . Most of them drive there.*

After "taking ground," the PSIU agents arrive at the scene: a dark alley with several police cars, municipal officials, and curious residents. There is a red car parked with the trunk open (similar to the one shown in Figure 6.1).

> POLICE OFFICER 1: *Who's the owner of this car?*
> A MAN IN HIS THIRTIES STEPS FORWARD: *Me.*
> POLICE OFFICER 1: *Turn it on.*

The car owner turns on the sound. A funk carioca song comes softly from the car speakers.

> POLICE OFFICE1 1: *Louder!*
> CAR OWNER: *That's the maximum it goes.*

Another police officer comes closer.

> POLICE OFFICER 2: *There is usually a button that activates the speakers in the trunk.*

He orders the car owner to get out of the car. After searching on the control panel, the police officer finds a hidden button under the steering wheel. It activates the trunk speakers and a funk ostentação beat hits me full blast. The police officer looks at the PSIU agent, suggesting he could measure the sound now. The PSIU agent points the SLM at the car. After a few seconds, the device provides the piece of information they all wanted:

> PSIU AGENT: *99 decibels.* [See Figures 6.4 and 6.5.]

The police officers gather around to see the number on the SLM screen. The PSIU agent is asked to take the fine form to the military police mobile base, where a police officer is checking the driver's record on a laptop.

> Police officer 3 [talking to the officer on the laptop]—*So how much is the fine?*
> POLICE OFFICER 4 [ON THE LAPTOP]: *I think it's R$191 [US$53], because it's not grave. That's very little for the disorder it causes.* [He reads Article 228 from the 1997 Brazilian Traffic Code] *"Use on the vehicle of sound equipment in the vehicle at a volume or frequency not authorized by the CONTRAN."*
> PSIU AGENT: *Is it possible to apprehend the vehicle?*
> POLICE OFFICER 4: *No. They apprehend for other reasons. . . . because of some irregularity with the driver's license. But in this case, everything is O.K. I'll see if I can find something to increase the punishment. There's always something else. Because this punishment here is too light—we can't take the car.*

The noise measurement is a problematic piece of evidence, you might say. In his defense, a PSIU inspector explained to me that "nobody would buy powerful car audio and not play it loud." At first, the PSIU inspectors would feel somewhat lost in these large operations, unsure of where to go, whom to talk to, and what to do. The PSIU agent who went on the first operation admitted that he was not sure how to measure the sound exactly. "A police officer told me it was *this* resolution from *this* article from the Brazilian Traffic Code, so I filled in the technical report the best I could," he explained. "After that, I looked on the Internet what and how I'm supposed to do this."

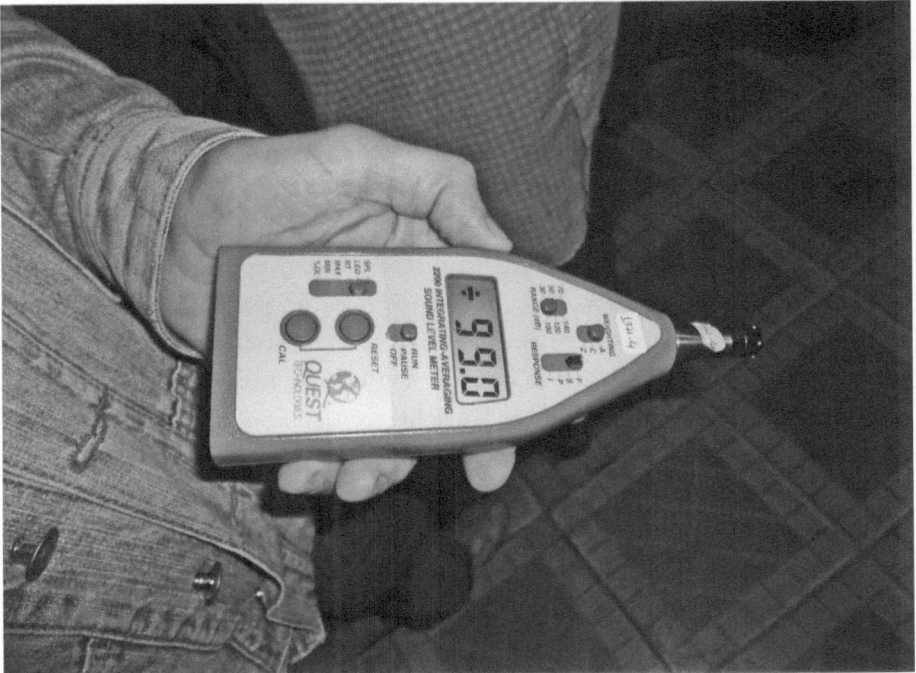

FIGURES 6.4 AND 6.5. PSIU agent fills the fine form (above) as other public officials try to see the numbers on the SLM screen (below).

Photo by the author, March 2012.

The theoretical (broken windows premise) and practical (legal channels and administrative flows) construction of a causal relation between "rowdy" teenagers and crime gave the police a legal "handle" from which they could act. Even when no illicit activity was found, the presence of loud sounds gave the state enough legal room to justify the police's forceful action. The M'Boi Mirim operation became a model for dealing with pancadões, and the Secretary of Subprefectures advised other subprefectures to coordinate similar operations. Every weekend, dark alleys in the poor peripheries became battlegrounds of partygoers against the police. Residents were acoustically squeezed between the boom of the pancadão car audio and the boom of the stun grenades thrown by the police. The operation showed to the communities that the state could act and establish order in the slums and suburbs too. For the police, the message was that zero-tolerance applied for the whole city. Unlike Rio de Janeiro, where several slums are heavily guarded by drug factions, in São Paulo the state can penetrate virtually anywhere, anytime (police officers made sure I understood this regional difference). The incorporation of nonlethal weapons has allowed the police to approach the public space with a zero-tolerance attitude without being accused of violence and abuse, as has happened so often in the past.

Rewiring the Echo Chamber

As the street parties continued spreading out and moving around, it seemed clear that the anti-pancadão operation would not be able to solve the problem. In 2011, Jooji Hato, the author of the 1:00 a.m. Law introduced in Chapter 3, was elected to the São Paulo State Chamber of Deputies. Besides proposing several bills related to alcohol consumption, he submitted Bill 924, to prohibit parked vehicles from playing music above 50 dB (measured at 6.5 feet and following the NBR10151) between 8:00 a.m. and 10:00 p.m., and at any volume between 10:00 p.m. and 8:00 a.m. Although it passed in the Chamber of Deputies, the governor ended up vetoing the bill, claiming that it was unconstitutional for the state to legislate on noise issues whose effects could only be determined at the city level. In 2011, the neighboring city of Osasco started to fine drivers US$3,200 for playing music above 70 dB from a car. Around that time, Diadema, a city at the forefront of dry laws in greater São Paulo,[33] added an article to its noise ordinance to fine car

---

[33] Diadema's Law 2107/2002 prohibits any venue that commercializes alcohol from opening from 11:00 p.m. to 6:00 a.m. More severe than Hato's 1:00 a.m. Law, this law also built on research suggesting a strong relation between alcohol consumption, nightlife activity, and homicide.

owners based on the difference in decibels between the sounds coming from the car and the limits established in zoning laws.

In 2012, Camilo (the former military police chief) was elected to the São Paulo Municipal Chamber. Along with Camilo, two other high-ranking police officers got elected as well. Because of their background and zero tolerance perspective, the press nicknamed the trio the Chamber's "Bullet Caucus" (*Bancada da Bala*). The Bullet Caucus quickly moved to try to extinguish the pancadão.[34] One bill, which passed into law in 2013,[35] includes a US$280 fine for anyone playing loud music in parked vehicles above the limits established by the zoning laws.[36] It also authorizes the confiscation of the car or the car audio equipment if the owner refuses to turn down the volume. This law was delegated to the PSIU.

The other document issued by the Bullet Caucus, Bill 2/2013, determines that "it is expressly forbidden to use streets, plazas, parks, and other public spaces for 'funk parties,' or for any nonauthorized musical event, regardless of duration." As with the other bill, the sanction includes the confiscation of the vehicle. The fine is US$320 for the first offense and doubles with each recurrence. Conte Lopes, the co-author of this bill with Camilo (and the second member of the Bullet Caucus), explained that "Public disorder will not end without a specific ordinance and until we have more rigorous policing" (personal communication, March 2013). For Lopes, who claimed there were four hundred pancadão points in São Paulo, the bill is a necessary response to "desperate residents, from various districts of São Paulo, begging for action." Workers' Party mayor Fernando Haddad vetoed the bill, arguing that it would affect negatively several municipal events that use the public space.

The pancadão was further engulfed into a broader debate on legal minority age. Lawmakers, the press, and public security officials started to vehemently challenge the extent to which the state should hold minors legally accountable for their acts. A survey conducted by *Folha de São Paulo* in 2013 showed that 93% of Paulistanos wanted the legal age reduced from eighteen to sixteen years old, and 35% supported reducing it to thirteen years old.[37] The "disorderly" use of public space has been a crucial element in framing the "undisciplined" youth as an enduring problem. As the comments from the CONSEG meetings suggest, although the debate relates implicitly to race, class, and gender, it is the generational framework that is used here; it has proven to be an efficient discursive mechanism for

---

[34] Camilo, who had left the police force to take office as newly elected councilman, co-authored both bills.
[35] Law 15777/2013, regulated by Decree 54734/2013.
[36] Re-incidence within thirty days doubles the fine, continuous re-incidences quadruples it.
[37] "Em SP, 93% apoiam redução da maioridade," *Folha de São Paulo* (April 28, 2013), C10.

politicians and public officials to avoid being accused of discrimination while hitting a political nerve firmly established around family, property, and taste. In 2013, a São Paulo councilman aligned with youth social movements submitted a bill for the creation of specific spaces for the pancadão. The idea of this bill was to take the youth away from the streets, decreasing noise complaints and concerns with violence and drug use. In 2014, the mayor vetoed the bill, suggesting that it would be wrong to privilege funk carioca fans exclusively by supporting parties with taxpayers' money. In 2015, the Brazilian Chamber of Deputies voted and approved in two separates sessions Bill 171/1993, a constitutional amendment that lowers the legal age from eighteen to sixteen years old. For Paulo Telhada, a reformed military police colonel and the third member of the Bullet Caucus in the São Paulo Municipal Chamber, the legal age should be reduced to fourteen years. "Here in São Paulo, fourteen-, fifteen-year-old criminals rape and kill, knowing what they are doing," Telhada explains. "Gangs use the minors because they know that if they are arrested, the minors assume the responsibility and stay incarcerated for one or two years at the most."[38]

At the core of this controversy is an ambiguous premise about the pancadão youth. On the one hand, the poor suburban youth are, like any other adolescent in the city, "people under development." They are not full-fledged citizens yet and thus cannot make decisions about their own lives. On the other hand, they need to be punished similarly to adults for their lack of discipline. The pancadão controversy and the zero-tolerance approach are thus tied to a critical question about citizenship: the disjuncture of deeds and rights. Bannister et al. argue in their analysis of zero tolerance policies like the Respect Action Plan in the United Kingdom that such measures risk stimulating "'criminalization of social policy' in that the focus of policy attention is placed upon behaviors and activities of certain groups rather the underlying socioeconomic causes of distress" (Bannister et al. 2006, 925).

Rewiring Ears and Norms

Finally, we come back to ears and norms. In "Loud Car Stereos," a report issued by the U.S. Department of Justice to assist police departments across the country in dealing with loud car audio, Michael S. Scott explains that "Low-frequency noise is usually found to be more annoying than high-frequency noise at similar volume. The vibrations caused by the low-frequency sound waves can often be felt

---

[38] "Deputado Cel. Telhada diz que 'infelizmente' matar 'faz parte da ação policial,'" *BBC Brasil* (August 24, 2015), http://www.bbc.com/portuguese/noticias/2015/08/150820_telhada_ping_jc_lk.

in addition to being heard. They cause glass and ceramics to rattle, compounding the annoyance" (Scott 2004, 2). In our analysis of Ears 1.0 and 2.0 from Chapter 2, we saw that most SLMs include "A" and "C" weightings (based on equal-loudness contours), where the former is modeled on the 40-phon curve (40 dB at 1 KHz), the latter on the 100-phon curve.

The perception of the pancadão music is sonic *and* tactile, making it particularly annoying for those included in this ad hoc party environment against their will. The SLM standard "A" weighting is not ideal for measuring the 20–100 Hz low-frequency spectrum. A typical funk ostentação song makes extensive use of the prolonged bass notes in the 50–100 Hz range (see Figure 6.6). Measuring a pancadão car playing this tune using the "A" weighting filter would decrease the result by 30 dB in relation to the 1 kHz 0 dB reference. If the PSIU inspector were to measure the car sound applying the "C" weighting filter, he would diminish only 1 dB approximately. In other words, between the two weightings, we have a nonnegligible difference of 29 dB of groovy bass. ISO 1996-1 (2003)[39] includes an informative annex addressing low-frequency sounds, which "engender greater annoyance than is predicted by the A-weighted sound pressure level" (ISO 2003, 18).[40] One way to deal with that, the document suggests, is to use octave-band or one-third-octave band analysis—which brings us to the debate mentioned in Chapter 2 about the costs and benefits of using more sophisticated and expensive SLMs.

As street parties with car audio spread out in the country, and since virtually no police departments are equipped with SLMs, the National Traffic Council revised the approach to Article 228 from the Brazilian Traffic Code. Resolution 204/2006 was replaced by Resolution 624/2016, which prohibits "peace disturbance [. . .] from any vehicle caused by equipment that generates sound audible from the exterior [of the vehicle], *regardless of volume or frequency* (emphasis mine)." The revised resolution thus gives the police enough authority to bypass Ears 1.0 and 2.0 to punish car owners and create a smoother punishment channel.

Let us recap. We have left and re-entered the sphere of state control with the pancadão. The youth articulate pancadões, using new (funk ostentação, mobile Internet) and old (car audio, streets) actors to taste loud grooves. The sound reaches the residents, who demand action from the state, clogging the military police call centers and flooding the CONSEG meetings claiming this is a *collective* problem

---

[39] This standard has been revised in 2016.
[40] In *Guidelines for Community Noise*, Berglund et al. argue, "A-weighted measures have been particularly criticized as not being accurate indicators of the disturbing effects of noises with strong low-frequency components [. . .]. However, these differences in prediction accuracy are usually smaller than the variability of responses among groups of people [. . .]. Thus, in practical situations the limitations of A-weighted measures may not be so important" (Berglund et al. 1999, 28).

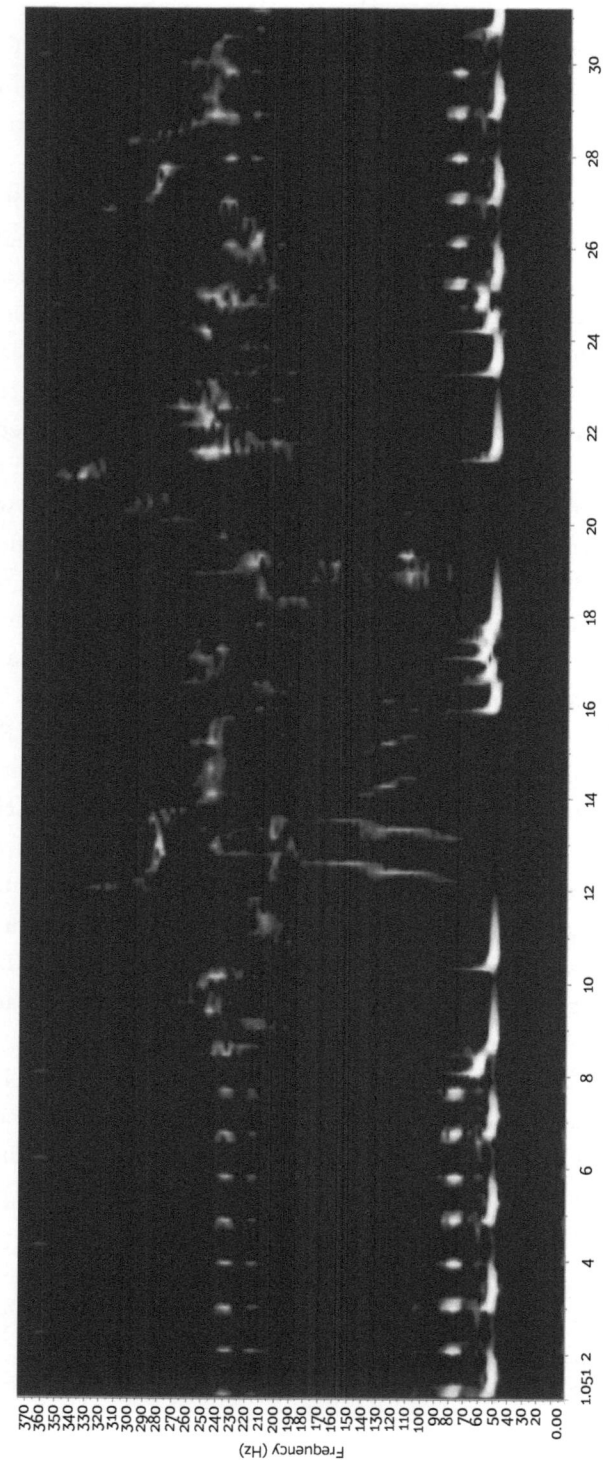

FIGURE 6.6. A spectrograph of an excerpt of MC Guimê's "Plaque de 100" ("$100 Bills") showing the prolonged bass notes in the 50–100 Hz range (for subwoofer performance). YouTube music video directed by Kondzilla available at https://www.youtube.com/watch?v=gyXkaOoDxB8.

that the state needs to solve. The police, who receive most of the complaints and are already frustrated with all the noise clogging their administrative flows, use a new actor (nonlethal weapons) to "gain ground" and restore quiet. The logic, which we encountered on several occasions throughout this book, is to tame noise to prevent "more serious crime." They collect the pertinent data and mobilize other state entities to make sure the operation has as few legal and bureaucratic gaps as possible. However, without the necessary personnel and structure to contain the hundreds of pancadões taking place every week, the police argue that the law is too permissive. Residents then reenter the controversy as constituents, pressuring the Chamber to include in its political circles the delegation of more authority to the executive branch. They elect lawmakers (many of them with police background) who will be willing to push for more discipline and punishment. Bills are created and passed, swinging the center of attention back to law enforcement.

In zigzagging across governmental agencies, from street to complaint, from office to fine, from car speaker to SLM, this chapter showed how the networks, considered in isolation earlier in the book, are in fact constantly in movement, affecting each other and calling for rewiring initiatives. The pancadão controversy made tangible the very notion of sound-politics as a conjuncture where groups and sounds are put into test and potentially reframed. It also allowed us to see the interesting and fluid relations between music and noise. Music mobilizes a passionate youth, who seek loud sounds to rewire their bodies collectively, to inscribe their identity in a democratic and inclusive space (the street), and to resist a moral mapping of citizenship that frames them as failed subjects and unfinished modern projects. Noise mobilizes the state, always concerned with disruption of property use and misuse of the same democratic and inclusive public space. In this oscillation, sound (as noise, music, and silence) remains the slippery object demanding the constant rewiring of chains of reference, political circles, administrative flows, and legal channels.

> In a time of so many crises in what it means to belong, the task of cohabitation should no longer be simplified too much. So many other entities are now knocking on the door of our collectives. Is it absurd to want to retool our disciplines to become sensitive again to the noise they make and to try to find a place for them?
> LATOUR, *Reassembling the Social* (2005, 262)

CONCLUSION

The Four Strata

IN CLOSELY EXAMINING how sounds enter and leave the sphere of state control, I attempted to delineate "noise" as a series of gestures rather than a fixed state of affairs. These gestures entailed the back-and-forth exchanges between state and civil society, noisemaker and noise complainant, sound source and human ears, Ears 1.0 and 2.0, and between governmental branches, office desks, institutional procedures, documents, inspections, maps, streets, sounds, and politics. My argument was that, at any point during their generation, propagation, circulation, reception, measurement, and elimination, sounds are subjected to a wide range of negotiations. It is the outcome of these negotiations that gives the heterogeneous noise some ontological stability.

Sound is sound-politics to the extent that it gets entangled with the inevitable challenge of cohabitation. Drawing on Latour, I addressed the hyphen in sound-politics by following controversies within different modes of existence (science, technology, politics, and law in particular). We saw that groups pertaining to a certain mode of existence often hear other modes and the discontinuities in their own mode as *noise*—unwanted, useless, unrefined, unreasonable, and unlawful noise. Rather than searching for consensus and smooth translations, the book examined a series of discontinuities and challenges created by the parasitic and multiple noise.

Conclusion 205

We have identified some of the most controversial sounds in São Paulo and considered how government actors articulate mechanisms to regulate these sounds. Having compiled this information, in this conclusion I summarize the findings presented in the book and suggest potential lines of inquiry for a study of citizenship and modernity. I do this by further elaborating on the notion of sound-politics, examining how sounds have entered and left the sphere of state control in São Paulo. Integrating the book's two analytical threads presented in the Introduction (modes of existence and governmental arrangements), I describe sound-politics in São Paulo as a field comprised of four strata: sonic complexes, debate axes, governmental dilemmas, and governmental solutions.

Stratum A: Sonic Complexes

Seven sonic complexes, each of which includes a collection of distinguishable sounds, were regularly encountered in this book: traffic, construction, industrial, religious, nightlife, street, and residential. In Chapter 1, I suggested that we subdivide the *traffic sonic complex* in terms of ad hoc behavioral sonic practices and larger, long-term, infrastructural sounds. The former ranges from car honking (which helped galvanize the first anti-noise waves in the city) to the more recent motorcycle explosive exhaust pipes mentioned in Chapter 3. The latter includes major controversies that made the notion of noise as *pollution* in the city tangible, especially the Congonhas Airport and the *minhocão* expressway.

Since the 1930s, concerns with street traffic congestion have become central to the very idea of urban planning in São Paulo. The dominance of the vehicle as the main framework to measure urban growth was stimulated by the car manufacturing and oil production industries, both of which had close ties with the Brazilian government (when they were not part of the government itself). This state-market alliance made sure that public problems such as noise did not interfere with growth in the automotive industry.

Despite its omnipresence, the traffic sonic complex has been mostly absent from state regulatory measures. Moreover, it is the parameter *against* which other sound sources are measured and fined. As a result, different branches of government have separately helped highlight this intriguing absence of state regulation for one of the loudest and most constant urban sounds. With no specific active technical standard targeting traffic noise, municipal lawmakers have been unsure as to how to deal with it. With no legal parameters, no administrative flows can be designed. With no administrative flows, no legal channels can be consistently built. Only in the early 2010s would lawmakers successfully push for a more proactive take on traffic noise control, in conjunction with the lobby from ProAcústica. It still, however, remains to be seen whether the noise map will become a reality, and whether the city administration will use it for future urban planning.

The *construction sonic complex* has also had a special status in São Paulo. Due to the city's rapid and extensive urbanization between the 1920s and 1970s, many of Brazil's major construction companies are headquartered in São Paulo. As with the traffic sonic complex, there are several alliances between private and public institutions here that have partitioned the city in order to maximize profit through the manipulation of pockets of land speculation along the South-West axis. For politicians such as Paulo Maluf (the mayor who built the minhocão), construction projects became part of the spectacle of urban modernity; they are proof of a visionary government that can deliver progress through grandiose monuments of steel and concrete.

As suggested in Chapter 1, the construction sonic complex has two main forms: the noise from construction work (public and private), which take place all day throughout the city; and as part of the soundproofing industry. Chapter 2 suggested that the centrality of civil construction to the creation of techno-scientific institutions such as the IPT (Technical Research Institute) and the ABNT has affected the regulation of construction noise in Brazil. The NBR 10151 and 1052 technical standards were created, and remain attached, to the ABNT Civil Construction Committee, which closely monitors their revisions and potential impact on "urban development." This became particularly apparent when the experts attempted to transfer the two technical standards to the newly created Acoustics Studies Committee.

The other side of the coin, the soundproofing market, started to boom in the 1950s with Eucatex thanks in part to the lack of regulation of both traffic and construction noise. Soundproofing companies began to envision the possibility of expanding their market by mobilizing some key players in the construction sector and strategically denouncing the ways that the state had protected traffic (noise). This alliance between construction and soundproofing (manufacturers and acousticians) was explicit, for instance, when Davi Akkerman (then president of ProAcústica) claimed that it was unjust for citizens and construction companies in Brazil to have to cover all the costs of environmental noise in the city while the city administration and state-regulated traffic were off the hook in the technical standards. Here we see the issue of territoriality surfacing as a matter of accountability in the tension between the private and the public.

Unlike the former two controversies, the *industrial sonic complex* has been more effectively regulated. As with most cities that industrialized in the early twentieth century, the distribution of factories in São Paulo was related to class-based geography, with working-class neighborhoods forming around factories and lower lands, and highly protected upscale districts built away from industry's environmental effects. The 1955 noise legislation discussed in Chapter 1 suggests that the city administration was already addressing industrial noise at that time. In the late 1960s, concerns about pollution and ecological sustainability helped lead to a sharper demarcation of the industrial sector in the city. There are many reasons why this controversy was less present during my fieldwork than others: factories are stationary noise sources; they were targeted early on by zoning legislation; there is a state agency (CETESB) dedicated to this type of noise; the industry produces different types of pollution (air, water, soil, and noise) at once making it an easy target of environmental regulation; and São Paulo has been going through a process of de-industrialization (Caldeira 2000).

Since the 1990s, the *religious sonic complex* has been at the center of noise legislation in São Paulo. As I showed in Chapter 3, this controversy is closely related to debates about religious freedom and concerns with the increase in Evangelical churches across the country. If complaints against Catholic churches and their regular bell-tolling are not uncommon, they pale in comparison to the thousands of complaints the PSIU receives every month against their Evangelical counterparts. The emphasis on "sonic excess" as an important performative element during the Evangelical services, and the impressive proliferation of churches (often in improvised garages) seem to amplify a sense of "sonic invasion," particularly among nonevangelical Paulistanos. Like traffic, construction, and industrial sonic complexes discussed so far, religious noise has direct ties to city lawmakers. Evangelical councilors have worked relentlessly to prevent the state from curtailing the acoustic ecstasy that forms the core of their religious practices.

The *nightlife noise complex* has important differences from the sonic complexes described above. First, in contrast to traffic, construction, and industrial noise, it is associated with "leisure." As shown in Chapter 3, it was precisely this perception of nightlife, valued as the necessary break from the work time with which São Paulo is so closely associated, that led lawmakers to warn about the potential damage of regulating bar noise. But when anti-noise campaigners were able to galvanize votes in the Chamber to pass the bill, the nightlife sector changed its response by qualifying and quantifying the activity in terms of job creation. Despite these attempts, the nightlife sonic complex has often been attacked as waste and a sign of incivility. In the 1990s, with fear of crime escalating and the international popularity of zero-tolerance policies, this sonic complex became engulfed in debates about public security. Additionally, in contrast to the religious sonic complex, nightlife was attacked by Evangelical lawmakers as the moral contamination of neighborhoods and the seeds of dysfunctional families. Nightlife should be tamed, they argued, because it was part and parcel of acts of promiscuity, carnality, inebriation, intoxication, and reckless euphoria. In that sense, nightlife has been associated with a wide range of concerns about public life, including peace, security, rest, and civility.

Like the nightlife sonic complex, to which it is closely associated, the *street sonic complex* is also immersed in fear of crime and public security concerns. As we saw in Chapters 3, 4, and 6, the conversion of this complex into public security matters had a legal platform with the military police, the institution legally responsible for maintaining order in the public space. The street sonic complex is particularly revealing for sound-politics studies because it pulls the state in different directions. It requires the government to mediate sonic disagreements between citizens while maintaining a certain level of neutrality. How can the state justify the presence of certain loud sounds and attack others without undermining social and cultural rights? Focusing on one group more than others can easily lead one to assume neutrality is being thrown out the window. *Carnaval*, political protests, strikes, sports rivalries (often performed with the effusive use of firecrackers), holidays, and the *pancadão* all take place in public spaces and require governmental mediation and decisionmaking. Chapter 6 suggested that broad accusations of bias were less common against the city's black youth, a marginalized group feared by the middle classes and with virtually no representation in the Municipal Chamber and other governmental institutions.

Finally, the *residential noise complex* is the most distant from the state due to well-established liberal conventions in which the government is expected to refrain from interfering with property rights. As we saw in Chapter 5, complainants who take legal action against their neighbors, seeking financial and moral compensation, can count on the state to be a neutral arbiter rather than an ally. It is up to the complainant to gather sufficient evidence to establish the "materiality" necessary for winning the case. The lack of state penetration in residential property besides a few articles in the Brazilian Civil Code has stimulated the growth of "Neighboring Law" jurisprudences among legal specialists. Doutrinadores such as Carneiro have the task of tying together codes and laws, thereby increasing the cohesion of judicial decisions in what remains a highly unstable legal channel.

Stratum B: Axes of Debate

In Chapter 1, I narrated the close relation between nuisance and zoning, and showed how noise ended up becoming a point of reference for organizing land use. We also saw an important ontological shift taking place in the city in the 1950s, when legislators moved away from a definition of noise based on specific sonic complexes to one based on decibel values. Here, I specify four different "axes of debate" that groups in São Paulo have used to defend and attack the sonic complexes described above. I present them here in chronological order according to when they appeared in the book.

AXIS 1: NOISE-CULTURE

In Chapter 1, I suggested that newspaper contributors across the six anti-noise waves (1910s–2010s) had insisted that the cause of so much noise in the city was the lack of civility. Whether reading newspaper stories from the 1930s or discussing the causes of environmental noise with acoustic engineers and soundproofing manufacturers in the 2010s, there seemed to be an ingrained culturalist explanation for the sonic practices experienced in São Paulo. One rationale operates within broad national contrasts. Unlike Japan, the United States, Germany, and France, Brazil is noisy; Brazilians (still) lack the minimum requisites for peaceful and respectful collective life. Another common rationale in the cultural spectrum relates to regional contrasts. For instance, the idea of racial mixing, celebrated in 1930s Rio de Janeiro by anthropologist Gilberto Freyre (1946) and considered as a crucial aspect of Brazil's cultural identity, was received with less enthusiasm in São Paulo. In fact, Paulistanos would use this cultural distinction to establish a local version of "The Grasshopper and the Ant": on the one hand, the carnivalesque and sensual Rio (and Brazil's Northeast) that merely profits from tourism and entertainment; on the other hand, the workaholic and modern São Paulo, which in a matter of decades was able to become the country's economic powerhouse. As Barbara Weinstein shows, this narrative of Paulista exceptionalism was permeated by race-related contrasts with other regions in the country (Weinstein 2015).

A final rationale operates within the city space. For instance, middle-class Paulistanos in the 1920s saw Afro-Brazilians in overpopulated tenements and Italian immigrants in working-class districts as a loud presence in the city and a reminder of the civilizing work still to be done. "Culture" has been the template groups use to situate nightlife, residential, street, and religious sonic complexes within a fluid noise-signal spectrum. Culture is the mechanism they use to dismiss and co-opt the axes of debate discussed below to frame the discussion as a matter of community values.

At an even more circumspect level, the noise-culture axis emerges when members of a given community perceive the introduction of certain sounds as interference and invasion. The sound of *carnaval* samba groups and a range of other music acts sanctioned by the state inevitably overflowed the streets and arenas from where they originated, seeping into adjacent spaces. A range of popular music genres (such as 1990s Miami Bass, 2000s Houston hip-hop, and 2010s funk ostentação) engineer low-frequency grooves by creating fields of sonic synergy between the car trunk and the street. When residents accept loud samba but condemn loud funk ostentação and the pancadão, they seem to refer to this process of community estrangement. It is out of this delicate intra-communal network of repulsions and attractions that the musical vortexes of loud grooves examined in Chapter 6 emerge.

Those accused of engaging in noisy activities can also claim they have cultural rights to do so, arguing that framing their music preference as "noise" is yet another form of ingrained racism, intolerance, and ethnocentrism. We saw that groups in the religious sonic complex often operate in this axis as well. Evangelical leaders claim that the message of the church should exclude their sounds from the noise category. The churches' sonic excess belongs to the mission of permeating neighborhoods with faith and hope, with a culture based on family values and the strong connection between body and spirit.

AXIS 2: NOISE-HEALTH

This axis relates to the physiological effects of noise on the human body. The World Health Organization (WHO) has been a crucial actor, helping build a scientific framework that governments can use to tackle environmental noise. Chapter 2 focused on the construction of Ears 1.0 and 2.0, the two building blocks in the chains of reference linking an objectified body to an objectified sound. Attached to hundreds of millions of ears, laws, scientific journals, labs, and inscription devices, this is the most authoritative of the four axes. Here, noise (potentially) enters the sphere of biopolitics, where quantifying potential deaths and diseases becomes a necessary strategy of population management. In contrast to the noise-culture axis, this axis claims neutrality and universality, setting itself above "localized" interpretations of noise. Advocates for the noise-health axis insist it can penetrate any of the seven sonic complexes and become a point of reference for assessing each of them in an unbiased manner.

However, this axis is not immune to shifts once it enters hybrid forums. In Chapter 2, the specialists revising the technical standards tried to establish noise measurement parameters by articulating the chains of reference to determine how sound travels from a source and reaches an "average" potential ear (itself folded into the circuitry of the sound level meter). In that chapter, we saw other groups constantly trying to insert themselves into the technical standards procedures and premises belonging to other axes. We saw that the noise-health axis is not authoritative across all modes of existence. For example, Chapter 5 showed that judges at the São Paulo Court of Appeals were not persuaded by the chains of reference linking noise, pollution, and injury.

AXIS 3: NOISE-MARKET

This axis is quite multifaceted. It relates to the ability of the car-oil complex to conceal traffic noise in the ABNT technical standards and to the legal struggles between the Congonhas airport and nearby residents. It relates to how real estate companies could build so much with so little concern about the noise they produced, only to later turn silence into a commodity for selling property. It relates to the ways in which powerful nightlife actors, such as bars, nightclubs, and alcoholic beverage and tobacco companies, have attempted to push for less restrictive noise control laws and law enforcement, and to land ownership and the fear of property devaluation caused by noisy activities.

Time and time again, we saw that the city government is willing to turn a deaf ear to noises coming from activities they consider necessary to the economic prosperity of the city. But ever since the first anti-noise waves, groups have often questioned the balance between profit and noise. For one, since the 1950s specialists have argued that environmental noise has negative effects on the workplace. More recently, notions of "quality of life" and "sustainability," placed alongside urban concerns such as air and water pollution and recycling, have pushed for a change in the city's noise-market axis.

AXIS 4: NOISE-CRIME

It is hard to sufficiently emphasize the centrality of the fear of crime in everyday urban Brazil. Brazilians, especially those who use public space frequently, learn from an early age how to scan and monitor an environment according to a tacit understanding of risk. In navigating the nightlife, the streets, and residential sonic complexes, they use what Barry Truax defines as "listening-in-search" (the scanning of the acoustic environment in search of useful sounds) and "listening-in-readiness" (developing the ability to filter out familiar sounds as background, ready to identify subtle changes) (Truax 1984). Before distinguishing interesting from uninteresting sounds, urbanites use listen-in-search and listen-in-readiness to search for the potentially criminal and violent (for some, this includes police violence).

Chapters 3, 4, and 6 showed how the combination of the fear of crime and zero-tolerance policing approaches have transformed sound-politics in São Paulo. Those chapters described how both the street and nightlife sonic complexes have become deeply entrenched in strategies to decrease violent crime. The centrality of crime has been so crucial that even an agency responsible for administrative inspections such as the PSIU has ended up moving from the noise-health to the noise-crime axis. Because of well-established fears widely circulated in the press, the noise-crime axis often supersedes the noise-market and the noise-culture axes. This was exemplified by Hato's 1999 noise ordinance, which won popular approval even after the bar union claimed that the law would kill profits and jobs. It was also exemplified by the argument that pancadão was an avenue to crime rather than a cultural right. As suggested in Chapter 4, the crime-culture short circuit associated with the city's poor and blacks has been pervasive in Brazil. Since colonial times, the police have been summoned to monitor minorities, identify crime "hotspots," and dismantle spontaneous congregations of the poor and nonwhite in the public space.

Stratum C: Governmental Dilemmas

What we saw with noise ordinances, the idea of noise maps, and the push for more comprehensive technical standards was the negotiation between actors representing the noise-health, noise-market, and, to a lesser extent, the noise-culture axes. The propagation of sonic complexes and the circulation of axes of debate create a series of dilemmas for the state. For the most part, the book has focused on such governmental dilemmas. It followed these dilemmas across two main lines of inquiry: first, within the well-established separation of legislative, executive, and judiciary powers; and second, across a range of modes of existence (science, technology, politics, and law). Chapter 2 considered dilemmas pertaining to science and technology, analyzing a range of controversies related to measurability: how to build Ears 1.0 and 2.0, where to conduct measurements, what decibel values to use, what approach to use (passive or proactive?), and even what terminology to employ (should they get rid of the subjective "noise"?).

Chapter 3 examined some lawmaking dilemmas. Here the governmental dilemmas were mainly those of political circling: the encoding of the impossible balance between the total unity and the total dissolution of groups as they moved around specific issues. We saw the renewal of political circles with Tripoli's Noise Law, as Evangelical politicians struggled to move their religious sonic complex away from the noise-health debate to the more favorable noise-culture axis.

We also saw that Hato's 1:00 a.m. Law targeted the nightlife sonic complex and put in motion debates across the noise-culture, noise-health, noise-crime, and noise-market axes.

As enforcers of materialized political circles, the executive branch has the task of maintaining this state of affairs. But what institution should it deploy? How can the state deal with such a multifaceted problem? What administrative flows should be used and how can they be used effectively? In Chapter 4, I argued that legal-bureaucratic, budgetary, and party affiliation concerns make a definitive answer to such questions impossible, and the search for them fruitless. As the major disciplinary (with the authority to enforce the law) and punishment (capable of issuing fines) actor in a vast governmental network, the executive branch faces the daunting task of justifying every step in its administrative flows.

Chapter 5 described the dilemmas in the judiciary, including the question of which court should decide on noise litigation—is it an issue of public or private law? In exploring a range of legal channels, we saw how the construction of evidence can change, as each legal channel entails a specific method for gathering information. At every turn, the cause-effect links necessary for the judge to make a decision were challenged by the ephemerality of the event that triggered the litigation.

Stratum D: Governmental Solutions

The stability and duration of the proposed solutions depend on the interaction between sonic complexes, axes of debate, and governmental dilemmas, all moving parts that interfere with one another. Although the relations between the strata are hard to predict, the state is still required to *act*. To come up with potential solutions, the state needs to navigate across the different strata presented here. Like a complex, multi-sited and fluid hybrid forum, these strata are constantly generating overflows that compromise the governmental solutions and require further revisions.

In short, the book suggested that the state follows two procedures for solving noise controversies. One procedure was to determine whether noise was an intrinsic quality or a conditional feature of certain actors. Presuming that noise was an intrinsic condition often led to a straightforward solution, requiring the state to simply neutralize the source or separate it from the residents. For example, factories are intrinsically noisy. The solution for their noise necessarily involves separating them from the city residents with soundproofing and zoning. The fact that the city administration has not succeeded in protecting the residents from the Congonhas airport speaks volumes about governmental leniency, the sway of the noise-market axis, and the complexity of noise pollution inside legal channels.

Unlike intrinsically noisy actors, conditionally noisy actors are not necessarily noisy, but they *can* be. Solutions for conditionally noisy actors are less straightforward and costlier, as these actors require continuous monitoring. Both the PSIU and the military police are governmental devices that are accountable for monitoring and punishing a wide range of conditionally noisy sources. These institutions can provide solutions for some (but not all) sonic complexes. As we saw in Chapters 2 and 4, the decibel unit has become an authoritative device to establish whether a sound is noise (i.e., harmful) or not. The decibel helps to stabilize monitoring procedures, as well as the administrative flows and legal channels attached to them. However, as Carneiro explained in Chapter 5, dependency on the decibel has become a problem because sound wave amplitude cannot (and should not) alone determine the effects of noise exposure on health.

The first procedure for finding a governmental solution refers to the constancy and localization of actors, which are either intrinsically or conditionally noisy. The second procedure for solving a noise controversy involves either tackling the sound wave itself or *something else* related to it. One example of the two approaches would be the two noise ordinances analyzed in Chapter 3. While in Tripoli's Law, the center was noise itself, Hato's 1:00 a.m. Law was only indirectly related to it. Rather, as Hato explained in his book (Hato 2010), the center of the 1:00 a.m. Law was alcohol consumption and crime prevention. Subsequently, as the strategy of closing bars at 1:00 a.m. to prevent crime came under scrutiny, the city made noise itself the focus of the law. Another example, also from Chapter 3, was the series of negotiations that took place in the Municipal Chamber, in which noise controversy was used as political leverage for deliberating other non-noise-related bills.

These two procedures operate independently. One would expect that the government deals with noise directly in cases of intrinsically noisy actors. But that is not always the case; even when the problem is noise itself, the state can deal with it by addressing it indirectly. As an intrinsically noisy actor, the state cannot solve the problem of Congonhas by requesting that it soundproof the noise source (a more direct approach) as it can with most factories. It can only propose indirect solutions, such as changing operating hours, subsidizing soundproofed windows on the buildings nearby, and establishing zoning measures.

It is often unclear which procedures the state is deploying. How can we know whether the government is implying that certain groups are intrinsically noisy? How can we be sure about whether it is providing a solution that deals with the sound wave itself or simply deploying noise control as a means to other ends?

Analyzing how noise solutions are attached to axes of debate can help us identify whether noise or something else is the main target of a piece of legislation, law-enforcement decision, or a judicial decision. As I have shown in this book, the use of these two procedures for solving noise problems can shift depending on the city administration (different political parties tend to focus on different procedures), the institution, and even the public official in charge. Moreover, as the controversy over Hato's bill suggests, politicians often shift the parameters of governmental solutions as they circle around the issue. Figure 7.1 shows the four strata that mediate sound-politics in São Paulo.

As mentioned in the Introduction, my intention was not to make a comprehensive, conceptual, and universalistic claim about urban noise. The reason I am presenting these sound-politics strata in the conclusion is precisely to emphasize its ethnographic grounding. This is also an ANT premise with important political implications. As Latour notes, "instead of taking a reasonable position and imposing some order beforehand, ANT claims to be able to find order much better after having let the actors deploy the full range of controversies in which they are immersed" (Latour 2005, 23).

I defined sound-politics as the process through which sounds enter and leave the sphere of state control. Using the analysis presented here, we can now revisit that initial, somewhat generic, statement. We can now say that sound-politics is *the multiple interactions within and between sonic complexes, axes of debate, governmental dilemmas, and governmental solutions*. Although time and resources allowed me to touch on only a few of the interactions, I hope it was enough to persuade political scientists, anthropologists, ethnomusicologists, sound studies scholars, sociologists, legal scholars, and historians about how sound is a viable path to study the state, and vice versa.

This brings me, finally, to citizenship. James Holston's distinction between differentiated and insurgent citizenship focuses on the relation between the state and civil society rather than as a matter of specific institutional configurations or membership statuses. Differentiated citizenship, where groups consider "social differences that are *not* the basis of national membership—primarily differences of education, property, race, gender, and occupation—to distribute different categories of citizens" (Holston 2008, 7), is a pervasive facet of Brazilian modernity. We saw this type of citizenship, for instance, when examining the revision of the technical standards (where the construction and traffic sectors tried to avoid disciplinary action), debates in the Municipal Chamber (with the Evangelical lawmakers separating their sounds from noise), and the assumptions that those who partake in the pancadão are proto-criminals.

Insurgent citizenship emerged in the late 1970s with the combination of rural-urban migration, unregulated urbanization, community activism, and

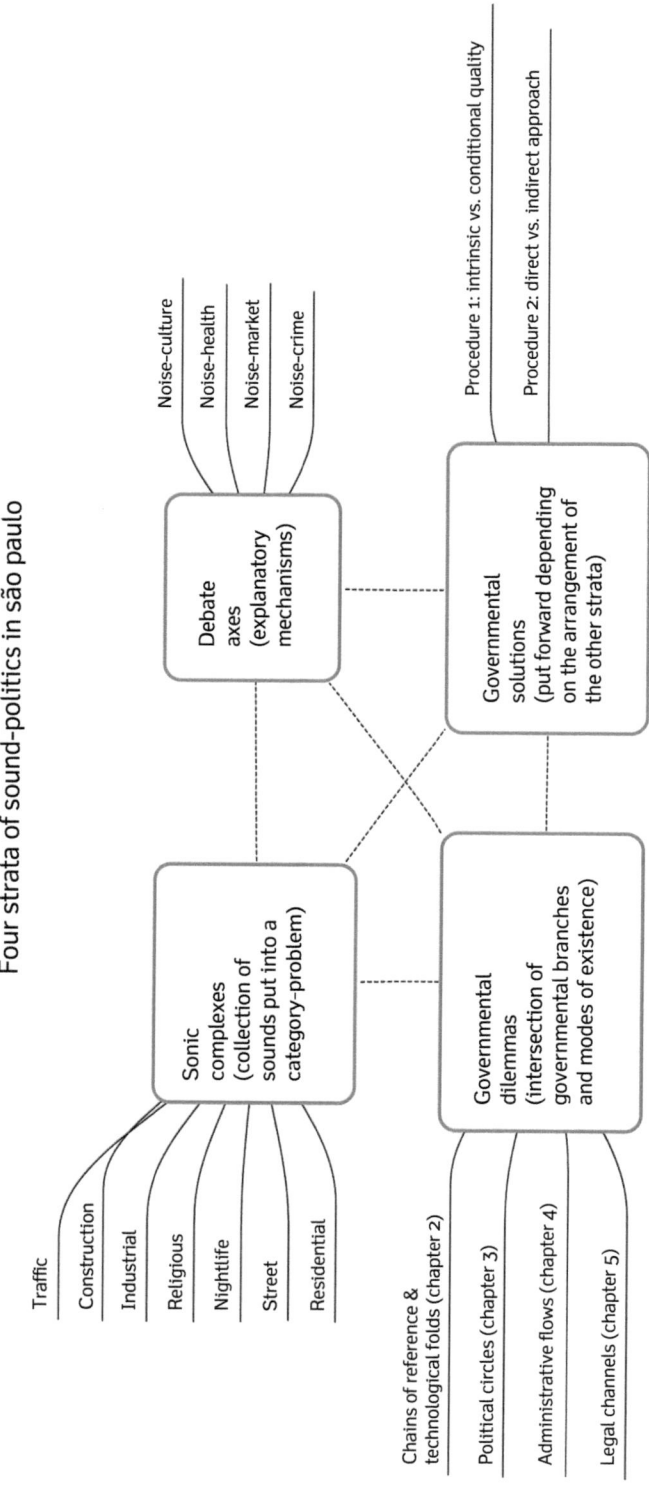

FIGURE 7.1. Four strata of sound-politics in São Paulo.

redemocratization. This model of citizenship is enacted when disenfranchised groups negotiate their symbolic and material incorporation to the formal city. We saw this with the street protest at the beginning of Chapter 3 and with the pancadão fans. Both, in their own ways, are pushing for a reformulation of what counts as a right to the urban space. One can argue that these insurgent groups are located at different levels of proximity to the encoded legal text, with the homeless movement mobilized around a cause (housing) through specific governmental actions (e.g., the vote on the City Master Plan), and the pancadão more loosely organized around a cause (the right to leisure) without well-defined spokespersons.

The pancadão controversy was rearticulated in 2013 when fans of funk ostentação started to get together outside the pancadão. They organized, via Facebook, a series of rolezinhos ("rolê" is slang for going out on the town) in shopping malls across the city. The shopping malls, which in Brazil are frequented by people across different classes and skin colors, were "invaded" by thousands of teenagers from the suburbs. The rolezinhos generated a moral panic in the press. The police responded with force and mall managers explained that they would bar any "suspicious" groups from entering their malls. The images of police officers forcefully searching for and expelling black teenagers from the malls underscored not only what some labeled the "Brazilian apartheid"[1] but also the unique place of funk carioca and its variants as markers of public disorder.

But more than describe these citizenship frameworks, all of which focus on groups claiming certain things from a supposedly unprejudiced and just state, I wanted to consider politics more broadly; not simply to talk about sounds as "markers of" disorder, culture, rest, or health, but as full-blown actors. Rather than claim that a sonic entity belongs to a specific sonic complex, axis of debate, governmental dilemma, or governmental solution, I used controversies to map out a range of possible sonic worlds. For example, in Chapter 4, rather than simply defining nightlife noise, I showed how actors negotiate different versions of it. If Ms. Freire's enactment of the noise she heard had failed to collaborate with the PSIU's version, or if the PSIU's version had not aligned with the ordinance's specifications, or if the bar owner somehow had managed to show the appellate judge different evidence, we would have ended up with an unsurpassable disjuncture between at least six objects: the sound, the noise complaint, the noise captured by the SLM, the noise fine, the noise ordinance, and the consequent sentence. It is the alignment of the object presented by Ms. Freire (a stance toward the

---

[1] "Rolezinhos: The Flash Mobs Currently Freaking Out Brazilian Authorities," *CityLab* (January 14, 2014). https://www.citylab.com/equity/2014/01/rolezinhos-flash-mobs-currently-freaking-out-brazilian-authorities/8130/.

bar sound) and by the state (the result of administrative flows and legal channels) that will determine whether she is going to have a restful night.

Here, I followed ANT's diplomatic approach to social analysis as a comparative ontology. Multiplying the number of sonic complexes by the number of axes of debate, the number of governmental dilemmas, and the number of governmental solutions, while considering the ontological specificities (the "keys") of modes of existence, one can get a sense of the numerous entities that could potentially be heard in a single noise controversy. It is in the arrangements of the four strata that we find not only noise's parasitic and multiple nature, but also the juncture of the book's two analytical threads.

I approached sounds as entities enmeshed in modes of existence *and* governmental branches, and presented sound-politics according to specific validation protocols, each of which turned (or attempted to turn) sound into a specific entity—scientific data, legal evidence, political asset, religious passion, administrative course of action, and so on. In discussing these protocols, I showed that there is more to social analysis than culture and capital. I am obviously not denying the relevance of economics in governmental, personal, and institutional decisionmaking—but I suggested that chains of reference, political circles, administrative flows, and legal channels deal with the pressure of capitalization in specific ways. I do claim, however, that the diplomatic approach to the modern institutions suggested by ANT allows for a richer and more nuanced analysis of the state.

To the São Paulo public officials, planners, lawmakers, law enforcers, judges, and civil groups, I suggest taking the four strata of sound-politics presented here as a starting point to tackle noise in the city. If noise is multiple as I have argued throughout the book, then the first step is to recognize its deeply relational nature. This means that singling out one or two sonic complexes, debate axis, or governmental dilemmas might not be the best strategy for finding a solution. The elements of each stratum should be considered together because they are constantly affecting each other. The second step is to make sure the actors brought to the negotiating table can understand each other without taking for granted their own claim to veracity as above the other registers. "What we want," argues Latour, "is an institution that follows the trajectory of its own mode of existence without prejudging the rest, without insulting the others and without believing, either, that it is going to be able to last in the absence of any reprise, through simple inertia" (Latour 2013, 482). Isabelle Stengers refers to this diplomatic project as "cosmopolitics." According to Stengers, "As an ingredient of the term 'cosmopolitics,' the cosmos corresponds to no condition, establishes no requirement. It creates the question of possible nonhierarchical modes of coexistence among the ensemble of inventions of nonequivalence, among the diverging values and obligations through which the entangled existences that compose it are affirmed" (Stengers 2011, 355–356).

If I have chosen to engage in an anthropology of the moderns, it is because I believe recalling and resetting modernity (Latour 2016) is an urgent matter not only for the Global North but also in places like Brazil, where the pressure to *embrace* modernity is particularly tangible. There is no question that this comparative ontology needs to be further decentered.[2] The enterprise attempted here provides the first sketches of an anthropology of the moderns that focuses on the multiple and parasitic noise as it traverses a range of Western-centered institutions.[3]

That the modern project has considerable weight in a place like São Paulo should come as a no surprise. Infrastructural deficiency and governmental leniency in one of the ten wealthiest cities in the world seems to have caused some cognitive dissonance among many Paulistanos. At the same time that they seem to have reached modernity with cutting-edge gadgets, worn-out passports (from constant international travels), and a clear-cut separation of nature and culture (Latour 1993),[4] they still have to deal with water shortages, power outages, long lines at airports, traffic jams, crime, and . . . noise. The use of terms such as "Belindia" (the wealth of a Belgium *and* the misery of an India) and "Ingana" (the taxes of an England *and* the public service quality of a Ghana) provided by Brazilian economists Edmar Bacha and Antônio Delfim Neto, respectively, suggest how Brazilians perceive themselves clumsily entrenched (or stuck, depending on whom you ask) in the modernity project. Decentering and extending such an inquiry to places like São Paulo can show, for instance, the zones of resonance, failure, and desire among those eager to move *toward* the same modernity so often critiqued in the humanities. Whether we take it as an after-effect, waste, or as "merely" a matter of taste, noise is a fruitful avenue for such an inquiry. It is my hope that sound-politics, as both an analytical framework and a methodological project, can contribute to our understanding of sound and politics. In considering noise's parasitic nature and ontological multiplicity, this book opens up new ways of studying the government. At the same time, in considering how different governmental institutions attempt to grasp such a slippery target, this book opens up new avenues for studying sound.

---

[2] This decentering of Western categories has been insightfully articulated by anthropologists Eduardo Viveiros de Castro (1992, 2014), Marilyn Strathern (1980, 1988), and Anna Tsing (2016). See also Charbonnier et al. (2017).

[3] An inquiry into the acoustic ontologies among Evangelical devotees, pancadão enthusiasts, nightclub owners, and noise complainants remains to be done.

[4] On the dire relations between agrobusiness and indigenous peoples in Brazil, see Danowski and Viveiros de Castro (2017).

References

ABNT (Associação Brasileira de Normas Técnicas). 2011. *Historia da Normalização Brasileira*. Rio de Janeiro: ABNT.
Araújo, Yago. 2016. "Cristofobia: uma análise do discurso de meios de comunicação evangélicos sobre a 19a Parada do Orgulho Gay de São Paulo." *Congresso Brasileiro de Ciências da Comunicação*. https://www.dropbox.com/s/ukxb05zmsnd2bnl/R11-1450-1.pdf?dl=0
Arias, Enrique Desmond and Daniel M. Goldstein. 2010. "Violent Pluralism: Understanding the New Democracies of Latin America." In *Violent Democracies in Latin America*, edited by Enrique Desmond Arias and Daniel M. Goldstein, 1–34. Durham: Duke University Press.
Arns, Paulo Evaristo. 2001. *Brasil: Nunca Mais*. Petrópolis: Editora Vozes.
Ashford, Norman, H. P. Martin Stanton, Clifton A. Moore, Pierre Coutu, and John Beasley. 2012. *Airport Operations*. New York: McGraw Hill.
Attali, Jacques. 1985. *Noise: The Political Economy of Music*. Minneapolis: University of Minnesota Press.
Avritzer, Leonardo. 2007. *Urban Reform, Participation and the Right to the City in Brazil*. Sussex: Institute of Development Studies.
Azevedo Marques, José Manuel de. 1932. "O Sossêgo Público em Face do Direito." *Revista dos Tribunais* 83 (389): 3–18.
Bannister, Jon, Nick Fyfe, and Ade Kearns. 2006. "Respectable or Respectful? (In)civility and the City." *Urban Studies* 43 (5/6): 919–937.
Barstow, J. M. 1940. "Sound Measurement Objectives and Sound Level Meter Performance." *Journal of the Acoustical Society of America* 12 (1): 150–166.
Békésy, Georg G. and Walter A. Rosenblith. 1948. "The Early History of Hearing: Observations and Theories." *Journal of the Acoustical Society of America* 20 (6): 727–748.
Berglund, Birgitta, Thomas Lindvall, and Dietrich H. Schwela (eds.). 1999. *Guidelines for Community Noise*. Geneva: World Health Organization.

Bertolli Filho, Cláudio. 1986. *Epidemia e Sociedade: a Gripe Espanhola no Município de São Paulo*. Master's thesis, Faculdade de Filosofia, Letras e Ciências Humanas, Universidade de São Paulo.

Bijsteverld, Karin. 2008. *Mechanical Sound: Technology, Culture, and Public Problems of Noise in the 20th Century*. Cambridge: MIT Press.

Bijsterveld, Karin. 2010. "Acoustic Cocooning: How the Car Became a Place to Unwind." *Senses & Society* 5 (3): 189–211.

Bijsterveld, Karin and Trevor Pinch (eds.). 2004. "Sound Studies: New Technologies and Music." *Social Studies of Science* 34 (5): 635–648.

Bijsterveld, Karin and Trevor Pinch (eds.). 2012. *The Oxford Handbook of Sound Studies*. Oxford: Oxford University Press.

Biondi, Luigi. 2009. "A Greve Geral de 1917 em São Paulo e a Imigração Italiana: Novas Perspectivas." *Cadernos AEL* 15 (27): 263–306.

Bonduki, Nabil. 1998. *Origens da Habitação Social no Brasil: Arquitetura Moderna, Lei do Inquilinato e Difusão da Casa Própria*. São Paulo: Estação Moderna.

Boulos, Guilherme. 2014. "A Batalha do Plano Diretor." *Folha de São Paulo* (July 1). https://www.dropbox.com/s/fz6js8zg1g2vxdx/A%20batalha%20do%20Plano%20Diretor%20-%2001%3A07%3A2014%20-%20Guilherme%20Boulos%20-%20ex-colunistas%20-%20Folha%20de%20S.Paulo.pdf?dl=0.

Boutin, Aimée. 2015. *City of Noise: Sound and Nineteenth-Century Paris*. Urbana: University of Illinois Press.

Brancatelli, Rodrigo. 2009. "Heliópolis começa a virar bairro." *O Estado de São Paulo* (July 7). https://brasil.estadao.com.br/noticias/geral,heliopolis-comeca-a-virar-bairro,398918.

Bratton, William. 1998. "Crime Is Down in New York City: Blame the Police." In *Zero Tolerance: Policing a Free Society*, edited by Norma Dennis, 29–43. London: IEA Health and Welfare Unit.

Brelàz, Gabriela de and Mário Aquino Alves. 2013. "O Processo de Institucionalização da Participação na Câmara Municipal de São Paulo: uma Análise das Audiências Públicas do Orçamento (1990–2010). *Revista da Administração Pública* 47 (4): 803–826.

Brenner, Neil, Bob Jessop, Martin Jones, and Gordon MacLeod (eds.). 2003. *State/Space: A Reader*. Malden, MA: Blackwell.

Brenner, Neil, Peter Marcuse, and Margit Mayer (eds.). 2012. *Cities for People, Not for Profit: Critical Urban Theory and the Right to the City*. New York: Routledge.

Bresser Pereira, Luiz Carlos. 1996. "Da Administração Pública Burocrática à Gerencial." *Revista do Serviço Público* 120 (1): 7–40.

Bretas, Marcos Luiz and André Rosemberg. 2013. "A História da Polícia no Brasil: Balanço e Perspectivas." *Topoi* 14 (26): 162–173.

Brown, Edward F., E. B. Dennis, and Jean Henry. 1930. *City Noise: The Report of the Commission Appointed by Dr. Shirley W. Wynne, Commissioner of Health, to Study Noise in New York City and to Develop Means of Abating It*. New York: City of New York.

Bull, Michael and Les Back. 2003. *The Auditory Culture Reader*. Sensory Formations Series. Oxford: Berg.

Caldeira, Teresa. 2000. *City of Walls: Crime, Segregation, and Citizenship in São Paulo*. Berkeley: University of California Press.

Caldeira, Teresa. 2012. "Imprinting and Moving Around: New Visibilities and Configurations of Public Space in São Paulo." *Public Culture* 24 (2): 385–419.

Caldeira, Teresa and James Holston. 1999. "Democracy and Violence in Brazil." *Comparative Studies in Society and History* 41 (4): 691–729.
Callon, Michel. 1998. *The Laws of the Markets*. Oxford: Blackwell.
Callon, Michel. 2017. *L'Emprise des Marchés: Comprendre Leur Fonctionnement pour Pouvoir les Changer*. Paris: La Découverte.
Callon, Michel, Pierre Lascoumes, and Yannick Barthe. 2009. *Acting in an Uncertain World: An Essay on Technical Democracy*. Cambridge: MIT Press.
Campos, Candido and Nadia Somekh. 2008. *A Cidade Não Pode Parar: Planos Urbanísticos de São Paulo no Século XX*. São Paulo: Mackenzie.
Canclini, Nestor Garcia. 2001. *Consumidores e Cidadãos: Conflitos Multiculturais Da Globalização*. Rio de Janeiro: Editora UFRJ.
Cano, Wilson. 1981. *As Raízes da Concentração Industrial em São Paulo*. São Paulo: T. A. Queiroz.
Cardia, Nancy. 1999. *Pesquisa sobre Atitudes, Normas Culturais e Valores Em Relação à Violência Em 10 Capitais Brasileiras*. Brasília: Ministério da Justiça, Secretaria de Estado dos Direitos Humanos.
Cardoso, Leonardo. 2018b. "Brazilian Hip Hop in Three Scenes." In *The Oxford Handbook of Hip Hop*, edited by Justin D. Burton and Jason Lee Oakes. doi: 10.1093/oxfordhb/9780190281090.013.21
Carneiro, Waldir de Arruda Miranda. 2014. *Perturbações Sonoras nas Edificações Urbanas: Ruídos em Edifícios, Direito de Vizinhança, Responsabilidade do Construtor, Indenização*. Belo Horizonte, Del Rey.
Carvalho, José Murilo de. 2001. *Cidadania no Brasil: O Longo Caminho*. Rio de Janeiro: Civilização Brasileira.
Castor, Belmiro Valverde Jobim. 2004. *O Brasil não é Para Amadores: Estado, Governo e Burocracia na Terra do Jeitinho*. Curitiba: Travessa dos Editores.
Cavalieri Filho, Sérgio. 2004. *Programa de Responsabilidade Civil*. São Paulo: Malheiros.
Centner, Ryan. 2012. "Microcitizenships: Fractious Forms of Urban Belonging after Argentine Neoliberalism." *International Journal of Urban and Regional Research* 36 (2): 336–362.
Charbonnier, Pierre, Gildas Salmon, and Peter Skafish. (eds.). *Comparative Metaphysics: Ontology After Anthropology*. London: Rowman & Littlefield.
Chesluk, Benjamin. 2004. "'Visible Signs of a City out of Control': Community Policing in New York City." *Cultural Anthropology* 19 (2): 250–275.
Childress, Herb. 2004. "Teenagers, Territory and the Appropriation of Space." *Childhood* 11 (2): 195–205.
Clifford, James. 1986. "Introduction: Partial Truths." In *Writing Culture: The Poetics and Politics of Ethnography*, edited by James Clifford and George E. Marcus, 1–26. Berkeley: University of California Press.
Corbin, Alain. 1998. *Village Bells: The Culture of the Senses in the Nineteenth-Century French Countryside*. New York: Columbia University Press.
Corten, André. 1999. *Pentecostalism in Brazil: Emotion of the Poor and Theological Romanticism*. New York: St. Martin's Press.
Costa, Frederico Lustosa da. 2008. "Brasil: 200 Anos de Estado; 200 Anos de Administração Pública; 200 Anos de Reformas." *Revista de Administração Público* 42 (5): 829–874.
Costa Leite, Manoel Carlos da. 1962. *Manual das Contarvenções Penais*. São Paulo: Saraiva.
DaMatta, Roberto. 1991. *Carnivals, Rogues, and Heroes: An Interpretation of the Brazilian Dilemma*. Notre Dame: University of Notre Dame Press.

Dagnino, Evelina. 2005. "Meanings of Citizenship in Latin America." *IDS Working Paper*. Brighton: Institute of Development Studies.

Danowski, Déborah and Eduardo Viveiros de Castro. 2017. *The Ends of the World*. Malden: Polity.

Delgado, José Augusto. 2008. "Responsabilidade Civil por Dano Moral Ambiental." *Informativo Jurídico da Biblioteca Ministro Oscar Saraiva* 19 (1): 81–153.

Dibble, Ken. 1995. "Hearing Loss & Music." *Journal of the Audio Engineering Society* 43 (4): 251–266.

Diniz, Maria Helena. 1995. *Código Civil Anotado*. São Paulo: Saraiva.

DJ Mr. Mixx. 2013. *The Bass That Ate Miami: The Foundation*. DVD. Directed by Alex J. Weir. Houston: Dreamhouse Studios.

Domingos, Roney and Juliana Carpanez. 2009. "Banda larga clandestina substitui lan houses na periferia de SP." *G1* (June 17). https://www.dropbox.com/s/oyyac2vltbvs6fp/G1%20%3E%20Edi%C3%A7%C3%A3o%20S%C3%A3o%20Paulo%20-%20NOT%C3%8DCIAS%20-%20Banda%20larga%20clandestina%20substitui%20lan%20houses%20na%20periferia%20de%20SP.pdf?dl=0.

Douglas, Mary. 1966. *Purity and Danger: An Analysis of Concepts of Pollution and Taboo*. London: Routledge.

Erlmann, Veit. 2004a. "But What of the Ethnographic Ear? Anthropology, Sound, and the Senses." In *Hearing Cultures: Essays on Sound, Listening, and Modernity*, edited by Veit Erlmann, 1–20. Oxford: Berg.

Erlmann, Veit. (ed.). 2004b. *Hearing Cultures: Essays on Sound, Listening, and Modernity*. Oxford: Berg.

Erlmann, Veit. 2010. *Reason and Resonance: A History of Modern Aurality*. New York: Zone Books.

Faria, Regina Helena Martins de. 2007. *Em Nome da Ordem: a Constituição de Aparatos Policiais no Universo Luso-Brasileiro (Séculos XVIII e XIX)*. PhD diss., Universidade Federal de Pernambuco.

Feld, Steven. 1988. "Aesthetics as Iconicity of Style, or 'Lift-up-over Sounding': Getting into the Kaluli Groove." *Yearbook for Traditional Music* 20: 74–113.

Feldman, Sarah. 2005. *Planejamento e Zoneamento em São Paulo: 1947–1972*. São Paulo: EDUSP.

Feltran, Gabriel. 2007. "Vinte Anos Depois: A Construção Democrática Brasileira Vista da Periferia de São Paulo." *Lua Nova* 72: 83–114.

Fernandes, Heloisa Rodrigues. 1989. "Rondas à Cidade: Uma Coreografia do Poder." *Tempo Social* 1 (2): 121–134.

Ferragi, Cesar Alves. 2010. "Koban and the Institutionalization of Community Policing in São Paulo." *Journal of Social Science* 70: 25–51.

Fletcher, Harvey and Wilden A. Munson. 1933. "Loudness: Its Definition, Measurement and Calculation." *Journal of the Acoustical Society of American* 5: 82–108.

Fórum Brasileiro de Segurança Pública. 2014. *Anuário Brasileiro de Segurança Pública 2014*. São Paulo: Fórum Brasileiro de Segurança Pública.

Fórum Brasileiro de Segurança Pública. 2016. *Anuário Brasileiro de Segurança Pública 2016*. São Paulo: Fórum Brasileiro de Segurança Pública.

Fórum Brasileiro de Segurança Pública. 2017. *Anuário Brasileiro de Segurança Pública 2017*. São Paulo: Fórum Brasileiro de Segurança Pública.

Foucault, Michel. 2007. *Security, Territory, Population: Lectures at the Collège de France, 1977–1978*. New York: Palgrave Macillan.

Foucault, Michel. 2008. *The Birth of Biopolitics: Lectures at the Collège de France, 1978–1979*. New York: Palgrave Macmillan.

Free, Edward E. 1930. "Practical Methods of Noise Measurement." *Journal of the Acoustical Society of America* 2 (1): 18–29.

Freyre, Gilberto. 1946. *The Masters and the Slaves: A Study in the Development of Brazilian Civilization.* New York: Knopf.

Gaetani, Francisco and Blanca Heredia. 2002. "The Political Economy of Civil Service Reform in Brazil: The Cardoso Years." Document Prepared for the Red de Gestión y Transparencia del Diálogo Regional de Política del Banco Interamerican de Desarrollo.

Gilroy, Paul. 2001. "Driving While Black." In *Car Cultures*, edited by Daniel Miller, 81–104. Oxford: Berg.

Gomart, Emilie and Antoine Hennion. 1999. "A Sociology of Attachment: Music Amateurs, Drug Users." *The Sociological Review* 47 (S1): 220–247.

Goodman, Steve. 2010. *Sonic Warfare: Sound, Affect, and the Ecology of Fear.* Cambridge: MIT Press.

Grin, Eduardo José. 2015. "Construção e Desconstrução das Subprefeituras na Cidade de São Paulo no Governo Marta Suplicy." *Revista de Sociologia e Política* 23 (55): 119–145.

Grosner, Diana, Daniela Gomes, and Renato Meirelles (eds.). 2013. *Vozes da Nova Classe Média: Caderno 3.* Governo Federal, Secretaria de Assuntos Estratégicos. https://www.dropbox.com/s/j00we6sdou5blu2/3cadernovcm-versofinal-130502162129-phpapp02.pdf?dl=0.

Hadfield, Phil. 2007. *Bar Wars: Contesting the Night in Contemporary British Cities.* Oxford: Oxford University Press.

Harvey, David. 1989. *The Condition of Postmodernity: An Enquiry into the Origins of Cultural Change.* Cambridge: Blackwell.

Hato, Jooji. 2010. *Alcool: Vetor da Violência.* São Paulo: Ekilibrio.

Helmke, Gretchen and Steven Levitsky (eds.). 2006. *Informal Institutions & Democracy: Lessons from Latin America.* Baltimore, MD: Johns Hopkins University Press.

Henderson, Jason. 2006. "Secessionist Automobility: Racism, Anti-Urbanism, and the Politics of Automobility in Atlanta, Georgia." *International Journal of Urban and Regional Research* 30 (2): 293–307.

Henriques, Julian. 2008. *Sonic Bodies: Reggae Sound Systems, Performance Techniques, and Ways of Knowing.* New York: Continuum International Publishing Group.

Hilliard, John K. 2006 [1984]. "Early History of the Evolution of the Volume Indicator." *Audio Engineering Society.* https://www.dropbox.com/s/evo4qea13310xsp/hilliard_early-history-of-vi.pdf?dl=0.

Holloway, Thomas H. 1993. *Policing Rio de Janeiro: Repression and Resistance in a 19th-Century City.* Stanford: Stanford University Press.

Holston, James. 2001. "Urban Citizenship and Globalization." In *Global City-Regions: Trends, Theory, Policy*, edited by Allen J. Scott, 325–348. Oxford: Oxford University Press.

Holston, James. 2008. *Insurgent Citizenship: Disjunctions of Democracy and Modernity in Brazil.* Princeton: Princeton University Press.

Hull, Matthew S. 2012. *Government of Paper: The Materiality of Bureaucracy in Urban Pakistan.* Berkeley: University of California Press.

Ising, Hartmut and Barbara Kruppa. 2004. "Health Effects Caused by Noise: Evidence in the Literature from the Past 25 Years." *Noise Health* 6 (5): 5–13.

ISO (International Organization for Standardization). 1997. *Friendship Among Equals: Recollections from ISO's First Fifty Years.* Geneva: ISO.

ISO. 2003. *ISO 1996: Description, Measurement and Assessment of Environmental Noise—Part 1: Basic Quantities and Assessment Procedures.* Geneva: ISO.

ISO. 2014. "ISO in Brief: Great Things Happen when the World Agrees." Geneva: ISO, 2016.

Jouili, Jeanette S. and Annelies Moors. 2014. "Introduction: Islamic Sounds and the Politics of Listening." *Anthropological Qurterly* 87 (4): 977–988.

Kang, Jian. 2006. *Urban Sound Environment*. London: Taylor & Francis.

Keeling, Kara and Josh Kun. 2011. "Introduction: Listening to American Studies." *American Quarterly* 63 (3): 445–459.

Keil, Charles. 1987. "Participatory Discrepancies and the Power of Music." *Cultural Anthropology* 2 (3): 275–283.

Khoury, Yara Aun. 1981. *As Greves de 1917 em São Paulo*. São Paulo: Cortez.

Kowarick, Lucio. 1985. "The Pathways to Encounter: Reflections on the Social Struggle in São Paulo." *New Social Movements and the State in Latin America*, edited by David Slater, 73–93. Amsterdam: Centre for Latin American Research and Documentation.

LaBelle, Brandon. 2008. "Pump up the Bass: Rhythm, Cars, and Auditory Scaffolding." *Senses and Society* 3 (2): 187–203.

Larkin, Brian. 2014. "Techniques of Inattention: The Mediality of Loudspeakers in Nigeria." *Anthropological Quarterly* 87 (4): 989–1015.

Latour, Bruno. 1987. *Science in Action: How to Follow Scientists and Engineers Through Society*. Cambridge: Harvard University Press.

Latour, Bruno. 1993. *We Have Never Been Modern*. Cambridge, MA: Harvard University Press.

Latour, Bruno. 2005. *Reassembling the Social: An Introduction to Actor-Network Theory*. Oxford; New York: Oxford University Press.

Latour, Bruno. 2007. "Turning Around Politics: A Note on Gerard Vries' Paper." *Social Studies of Science* 37 (5): 811–820.

Latour, Bruno. 2010. *The Making of Law: An Ethnography of the Conseil d'Etat*. Malden: Polity Press.

Latour, Bruno. 2013. *An Inquiry into Modes of Existence: An Anthropology of the Moderns*. Cambridge: Harvard University Press.

Latour, Bruno. (ed.). 2016. *Reset Modernity!* Cambridge: MIT Press.

Latour, Bruno and Steve Woolgar. 1986. *Laboratory Life: The Construction of Scientific Facts*. Princeton: Princeton University Press.

Law, John. (ed.). 1986. *Power, Action, and Belief: A New Sociology of Knowledge*. Boston: Routledge & Kegan Paul.

Law, John. (ed.). 1991. *A Sociology of Monsters: Essays on Power, Technology, and Domination*. London: Routledge.

Lefebvre, Henri. 1991. *The Production of Space*. Oxford: Basil Blackwell.

Lefebvre, Henri. 1996. *Writings on Cities*. Oxford: Basil Blackwell.

Lima, Sérgio de, Samira Bueno, and Guaracy Mingardi. 2016. "Estado, Polícias e Segurança Pública no Brasil." *Revista Direito GV* 12 (1): 49–85.

Liverman, Diana M. and Silvina Vilas. 2006. "Neoliberalism and the Environment in Latin America." *Annual Review of Environment and Resources* 31: 327–363.

Machado, Nelson. 2002. *Sistema de Informação de Custo: Diretrizes para Integração ao Orçamento Público e à Contabilidade Governmental*. PhD diss., Universidade de São Paulo.

Mainwaring, Scott. 1999. *Rethinking Party Systems in the Third Wave of Democratization: the Case of Brazil*. Stanford: Stanford University Press.

Maitland, Frederic William. 1885. *Justice and Police*. London: Macmillan.

Manco, Tristan, Caleb Neelon, and Lost Art. 2005. *Graffiti Brasil*. London: Thames & Hudson.

Mangin, William. 1967. "Latin American Squatter Settlements: A Problem and a Solution." *Latin American Research Review* 2 (3): 65–98.

Mariano, Ricardo. 2004. "Expansão Pentecostal no Brasil: o Caso da Igreja Universal." *Estudos Avançados* 18 (52): 121–138.

Massumi, Brian. 2002. *Parables for the Virtual: Movement, Affect, Sensation*. Durham: Duke University Press.

Matless, David. 1995. "The Art of Right Living: Landscape and Citizenship, 1918–39." In *Mapping the Subject: Geographies of Cultural Transformation*, edited by Steve Pile and Nigel Thrift, 85–113. New York: Routledge.

Meirelles, Hely Lopes. 2016. *Direito Administrativo Brasileiro*. São Paulo: Malheiros.

Merryman, John Henry and Rogelio Pérez-Perdomo. 2007. *The Civil Law Tradition: an Introduction of the Legal Systems of Europe and Latin America*. Stanford: Stanford University Press.

Mills, Mara. 2011. "Deafening: Noise and the Engineering of Communication in the Telephone System." *Grey Room* 43: 118–143.

Miraglia, Paula. 2011. "Homicídios: guias para a interpretação da violência na cidade." In *São Paulo: Novos Percursos e Atores*, edited by Lúcio Kowarick and Eduardo Marques, 321–346. São Paulo: Editora 34.

Mol, Annemarie. 1999. "Ontological Politics: A Word and Some Questions." In *Actor-Network Theory and After*, edited by John Law and John Hassard, 74–89. Oxford: The Sociological Review.

Mol, Annemarie. 2002. *The Body Multiple: Ontology in Medical Practice*. Durham: Duke University Press.

Moreno, Gilberto Geribola. 2011. "Novinhas, Malandras e Cachorras: Jovens, Funk Sexualidade." *Ponto Urbe* 9. https://www.dropbox.com/s/hh1m5i9th3jyj94/pontourbe-277.pdf?dl=0.

Mouffe, Chantal. 2000. *The Democratic Paradox*. London: Verso.

Neri, Marcelo, Ricardo Paes de Barros, Diana Grosner, Rosane Mendonça, Adriana Mascarenhas, Andrezza Rosalém, and Samuel Franco. 2013. *Juventude Levada em Conta: Demografia*. Governo Federal, Secretaria de Assuntos Estratégicos. https://www.dropbox.com/s/zt9s6obepy381t3/IPEA_juventude_2013.pdf?dl=0.

New York City. 1923. *Regional Plan of New York and its Environs*. New York.

Novak, David. 2013. *Japanoise: Music at the Edge of Circulation*. Durham: Duke University Press.

Novak, David and Matt Sakakeeny (eds.). 2015. *Keywords in Sound*. Durham: Duke University Pres.

Ochoa Gautier, Ana María. 2006. "Sonic Transculturation, Epistemologies of Purification and the Aural Public Sphere in Latin America." *Social Identities* 12 (6): 803–825.

Ochoa Gautier, Ana María. 2014. *Aurality: Listening and Knowledge in Nineteenth-Century Colombia*. Durham: Duke University Press.

Oosterbaan, Martijn. 2009. "Sonic Supremacy: Sound, Space and Charisma in a Favela in Rio de Janeiro." *Critique of Anthropology* 29 (1): 81–104.

Ortega, Roque Roldán. 2004. *Models for Recognizing Indigenous Land Rights in Latin America*. Washington, DC: World Bank.

Osborne, David and Ted Gaebler. 1993. *Reinventing Government: How the Entrepreneurial Spirit is Transforming the Public Sector*. New York: Penguin Books.

Pereira, Alexandre Barbosa. 2010. "A Maior Zoeira": Expriências Juvenis Na Periferia de São Paulo. PhD diss., Universide de São Paulo.

Pereira, Carlos and Bernardo Mueller. 2002. "Comportamento Estratégico em Presidencialismo de Coalizão: as Relações entre Executivo e Legislativo na Elaboração do Orçamento Brasileiro." Revista de Ciências Sociais 45 (2): 265–301.

Peterson, Marina. 2017. "Atmospheric Sensibilities: Noise, Annoyance, and Indefinite Urbanism." Social Text 35 (2[131]): 69–90.

Phillips, Adam. 2010. On Balance. New York: Farrar, Straus, and Giroux.

Pollard, Charles. 1998. "Zero Tolerance: Short-Term Fix, Long-Term Liability?" In Zero Tolerance: Policing a Free Society, edited by Norma Dennis, 44–61. London: IEA Health and Welfare Unit.

Prefeitura de São Paulo. 2012. Relatório de Gestão 2012: Secretaria Municipal de Coordenação das Subprefeituras. São Paulo: Prefeitura de São Paulo.

Rabelo, Carina. 2015. "Criminalidade só diminui quando todos os delitos são punidos." Coronel Camilo's Website, June 17. https://www.dropbox.com/s/d2trk8my9sexc7y/Criminalidade%20s%C3%B3%20diminui%20quando%20todos%20os%20delitos%20s%C3%A3o%20punidos%20%7C%20Deputado%20Coronel%20Camilo%20%7C%20Seguran%C3%A7a%20P%C3%BAblica%20%7C%20S%C3%A3o%20Paulo.pdf?dl=0.

Radovac, Lilian. 2011. "'The' War on Noise': Sound and Space in La Guardia's New York." American Quarterly 63 (3): 733–760.

Rama, Ángel. 1996. The Lettered City. Durham: Duke University Press.

Rede Nossa São Paulo. 2016. "Mapa da Desigualdade." https://www.dropbox.com/s/pzajmjqfj2f967x/mapa-da-desigualdade-completo-2016.pdf?dl=0

Robinson, D. W. and R. S. Dadson. 1956. "A Re-Determination of the Equal-Loudness Relations for Pure Tones." British Journal of Applied Physics 7 (5): 166–181.

Rolnik, Raquel. 1997. A Cidade e a Lei: Legislação, Política Urbana e Territórios na Cidade de São Paulo. São Paulo: Livros Studio Nobel.

Roma, Celso. 2002. "A Institucionalização do PSDB entre 1988 e 1999." Revista Brasileira de Ciências Sociais 17 (49): 71–92.

Romero, Simon. 2013. "Bus-Fare Protests Hit Brazil's Two Biggest Cities." New York Times (June 13). http://www.nytimes.com/2013/06/14/world/americas/bus-fare-protests-hit-brazils-two-biggest-cities.html.

Rommen, Timothy. 2011. Funky Nassau: Roots, Routes, and Representation in Bahamian Popular Music. Berkeley: University of California Press.

Rose, Tricia. 1994. Black Noise: Rap Music and Black Culture in Contemporary America. Middletown: Wesleyan.

Rosemberg, André. 2010. De Chumbo e Festim: Uma História da Polícia Paulista no Final do Império. São Paulo: EDUSP.

Rossing, Thomas D. 2007. "A Brief History of Acoustics." In Springer Handbook of Acoustics, edited by Thomas D. Rossing, 9–24. New York: Springer.

Samuels, David W., Louise Meintjes, Ana Maria Ochoa, and Thomas Porcello. 2010a. "The Reorganization of the Sensory World." Annual Review of Anthropology 39: 51–66.

Samuels, David W., Louise Meintjes, Ana Maria Ochoa, and Thomas Porcello. 2010b. "Soundscape: Towards a Sounded Anthropology." Annual Review of Anthropology 39: 329–345.

Sandroni, Carlos. 2001. Feitiço Decente: Transformações do Samba no Rio de Janeiro, 1917–1933. Rio de Janeiro: Jorge Zahar Editor.

São Paulo City Hall Website. 2017. "Dados Demográficos dos Distritos Pertencentes às Prefeituras Regionais." https://www.dropbox.com/s/nbxy1cu4y6b4qpz/Dados%20

demogr%C3%A1ficos%20dos%20distritos%20pertencentes%20%C3%A0s%20 Prefeituras%20Regionais%20%7C%20Secretaria%20Municipal%20de%20Prefeituras%- 20Subprefeituras%20%7C%20Prefeitura%20da%20Cidade%20de%20S%C3%A3o%20Paulo. pdf?dl=0.

Sarig, Roni. 2007. *Third Coast: Outkast, Timbaland, and How Hip Hop Became a Southern Thing*. Cambridge, MA: Da Capo Press.

Schafer, R. Murray. 1994. *The Soundscape: Our Sonic Environment and the Tuning of the World*. Rochester, VT: Destiny Books.

Scott, H. H. 1957. "Historical Development of the Sound Level Meter." *Journal of the Acoustical Society of America* 29 (12): 1331–1333.

Scott, Michael S. 2004. *Loud Car Stereos*. U.S. Department of Justice, Office of Community Oriented Policing Services. https://www.dropbox.com/s/7q9m3vbbte8s84f/682616. pdf?dl=0.

Serres, Michel. 1982. *The Parasite*. Baltimore, MD: John Hopkins University Press.

Silva, Daniela Wosiack da, Selma Maffei de Andrade, Darli Antonio Soares, Elisabete de Fátima P. de Almeida Nunes, and Regina Melchior. 2008. "Condições de Trabalho e Riscos no Trânsito Urbano na Ótica de Trabalhadores Motociclistas." *Physis* 18 (2): 339–360.

Sirqueira, Natália Gonçalves. 2012. "Aplicação Processual da Imunidade Tributária dos Templos e Cultos Religiosos." *Boletim Jurídico*. https://www.dropbox.com/s/1bhz89lhkjortvf/ Aplica%C3%A7%C3%A3o%20processual%20da%20imunidade%20tribut%C3%A1ria%20 dos%20templos%20e%20cultos%20religiosos%20-%20Boletim%20Jur%C3%ADdico. pdf?dl=0.

Smith, Mark M. (ed.). 2004. *Hearing History: A Reader*. Athens: University of Georgia Press.

Smith, Mark M. 2006. *How Race Is Made: Slavery, Segregation, and the Senses*. Chapel Hill: University of North Caroline Press.

Stengers, Isabelle. 2011. *Cosmopolitics II*. Minneapolis: University of Minnesota Press.

Sterne, Jonathan. 2003. *The Audible Past: Cultural Origins of Sound Reproduction*. Duke University Press.

Sterne, Jonathan (ed.). 2012. *The Sound Studies Reader*. New York: Routledge.

Sterne, Jonathan. 2015. "Hearing." In *Keywords in Sound*, edited by David Novak and Matt Sakakeeny, 65–77. Durham: Duke University Press.

Strathern, Marilyn. 1980. "No Nature, No Culture: The Hagen Case." In *Nature, Culture and Gender*, edited by Carol Maccormack and Marilyn Strethertn, 174–222. Cambridge: Cambridge University Press.

Strathern, Marilyn. 1988. *The Gender of the Gift*. Berkeley: University of California Press.

Sullivan, J. L. 1971. "A Laboratory System for Measuring Loudness Loss of Telephone Connections." *Bell System Technical Journal* 50 (8): 2663–2739.

Suzuki, Yôiti and Hisashi Takeshima. 2004. "Equal-Loudness-Level Countours for Pure Tones." *Journal of the Acoustical Society of America* 116 (2): 918–933.

Thompson, Emily. 2002. *The Soundscape of Modernity: Architectural Acoustics and the Culture of Listening in America, 1900–1933*. Cambridge: MIT Press.

Todd, Neil P. McAngus and Frederick W. J. Cody. 2000. "Vestibular Responses to Loud Dance Music: A Physiological Basis of the 'Rock and Roll Threshold'?" *Journal of the Acoustical Society of America* 107 (1): 496–500.

Todd, Neil P. 2001. "Evidence for a Behavioral Significance of Saccular Acoustic Sensitivity in Humans." *Journal of the Acoustical Society of America* 110 (1): 380–390.

Torres, Marcelo Douglas de Figueiredo. 2004. *Estado, Democracia e Administração Pública no Brasil*. Rio de Janeiro: FGV Editora.

Tremlett, Giles. 2012. "Cuba Cracks Down on 'Vulgar' Reggaeton Music." *The Guardian* (December 6). https://www.theguardian.com/world/2012/dec/06/cuba-crackdown-vulgar-reggaeton-music.

Trevisan, Janine. 2013. "A Frente Parlamentar Evangélica: Força Política No Estado Laico Brasileiro." *Numen* 16 (1): 581–609.

Truax, Barry. 1984. *Acoustic Communication*. Norwood, NJ: Ablex Publishing Corporation.

Tsing, Anna. 2016. "Earth Stalked by Man." *Cambridge Journal of Anthropology* 34 (1): 2–16.

UNAS. N.d. "Heliópolis, Maior Favela de São Paulo." https://www.dropbox.com/s/0rp78ot7vhyr83i/UNAS%20Heli%C3%B3polis%20e%20Regi%C3%A3o%20%7C%20HELI%C3%93POLIS.pdf?dl=0.

Viveiros de Castro, Eduardo. 1992. *From the Enemy's Point of View*. Chicago: University of Chicago Press.

Viveiros de Castro, Eduardo. 2014. *Cannibal Metaphysics*. Minneapolis: Univocal.

Vogel, Donald A., Patricia A. McCarthy, Gene W. Bratt, and Carmen Brewer. 2007. "The Clinical Audiogram: Its History and Current Use." *Communication Disorders Review* 1 (2): 81–94.

Wagner, David. 2012. "Cuba Banned Reggaeton and People are Surprisingly OK with That." *The Atlantic* (December 7). https://www.theatlantic.com/international/archive/2012/12/cuba-bans-reggaeton/320723/.

Walker, Warren. 1941. *The Planning Function in Urban Government*. Chicago: University of Chicago Press.

Walser, Robert. 2014. *Running with the Devil: Power, Gender, and Madness in Heavy Metal Music*. Middletown: Wesleyan University Press.

Ward, Keith. 2006. "A Short History of Telecommunications Transmission in the UK." *Journal of the Communications Network* 5: 30–41.

Weiner, Isaac. 2014. *Religious Out Loud: Religious Sound, Public Space, and American Pluralism*. New York: New York University Press.

Weinstein, Barbara. 2015. *The Color of Modernity: São Paulo and the Making of Race and Nation in Brazil*. Durham: Duke University Press.

Westhoff, Ben. 2011. *Dirth South: OutKast, Lil Wayne, Soulja Boy, and the Southern Rappers who Reinvented Hip-Hop*. Chicago: Chicago Review Press.

Weyland, Kurt. 1998. "The Politics of Corruption in Latin America." *Journal of Democracy* 9 (2): 108–121.

WHO (World Health Organization). 2011. *Burden of Disease from Environmental Noise: Quantification of Healthy Life Years Lost in Europe*. Copenhagen: World Health Organization.

Willis, Graham. 2009. "Deadly Symbiosis? The PCC, the State, and the Institutionalization of Violence in São Paulo, Brazil." In *Youth Violence in Latin America*, edited by G. Jones, 167–181. New York: Palgrave Macmillan.

Wilson, James Q. and George L. Kelling. 1982. "Broken Windows: The Police and Neighborhood Safety." *Atlantic* (March). http://www.theatlantic.com/past/docs/politics/crime/windows.htm.

Wolfe, Joel. 2010. *Autos and Progress: The Brazilian Search for Modernity*. Oxford: Oxford University Press.

Wynne, Shirley W. 1930. "New York City's Noise Abatement Commission." *Journal of Acoustical Society of America* 2 (12): 12–17.

Yúdice, George. 2003. *The Expediency of Culture: Uses of Culture in the Global Era*. Durham: Duke University Press.

Zukin, Sharon. 1995. *The Cultures of Cities*. Cambridge: Blackwell.

Index

ACOEM Group, 60
Acoustical Society of America, 49–50, 51–52
acoustic comfort, 71
acousticians, 59–60
acoustic maps. *See* noise maps
acoustics, architectural, 48, 60
Acoustics Commission controversies
    borders, establishing technical vs. social, 72–73
    decibel values, periods, and rooms, 66–69
    measurement, 69–70
    present vs. absent receiver, 62–64
    scope and traffic, 64–66
    vocabulary changes, 70–72
Acoustics Commission participants
    classes of, 58
    construction sector, 73
    law enforcement agencies, 59
    noisemakers, 60–62
    product providers, 60
    scholars, 58
    service providers, 59–60
    technical institutions, 58
    traffic sector, 73
Acoustics Commission revised norms
    business expansion and, 59–60
    legal ramifications, interest in, 59
    noisemakers interest in, 60–61
    scientific accuracy, interest in, 58
Acoustics Commission Studies Committee, 61, 71–72
acoustics field, 26, 48
Acoustics Performance Studies Commission, 58
Acoustic Technical Committee, confirming ISO 226 (2003), 50–52
acoustimeter, 51–52
activity of exclusion, 2–3
actor-network theory (ANT), 10, 15–16, 170–71, 215, 218
Administrative Department of Public Service, 129
administrative flows
    CONSEG meetings, rewiring, 192–98
    introduction, 101–3
    Military Police Operations Center (*Centro de Operações da Polícia Militar*, COPOM), 106, 107
    pancadão rewiring, 192–98
    São Paulo Anti-Noise Agency (*Programa de Silêncio Urbano*, PSIU), 121–26, 127, 155–56

232 Index

advertisements, 120
Akkerman, Davi, 93, 170–71
alcohol consumption
  crime and, 91–92, 214
  at pancadãoes, 182
American Standards Association, 51–52
American Tentative Standards for Noise
    Measurements, 51–52
Anti-Noise Agency. *See* São Paulo Anti-
    Noise Agency (*Programa de Silêncio
    Urbano*, PSIU)
anti-noise technology, 6–7
anti-noise waves
  1910s, 20–25
  1930s, 25–29
  1950s, 29–34
  1970s, 34–40
  1990s, 40–41
  2010s, 40–41
  acoustic traits of uncivility, focus on, 21–24
  economics of, 20–21, 211
  health issues focus during, 25
  hygiene and safety regulations, 24, 25
  immigration and the, 20
  introduction, 11–12
  newspaper contributors to, 209
  summary overview, 41–44
  urbanization and densification and the, 20
anti-*pancadão* operations, 191–98
anti-traffic noise laws, 28–29
Antônio, João, 90–91, 92
Apolinário, Carlos, 85–87, 92–93
appeal, right of, 148
appellate judges, environmental noise
    litigation, 152–55
architectural acoustics, 48, 60
architectural noise, 39–40
"Assessment of Noise in Inhabited Areas,
    Seeking the Comfort of the
    Community" (NBR), 55
AtenuaSom, 60
attachments
  groove, 170–72
  ideological, 136
Attali, Jacques, 4
audioanalgesia, 2–3
audiogram, 50–51
audiometer, 50–52

audiomobility, 176–80, 182
auto crime, 183–84, 185–86
auto industry, 28
automobiles, music from parked, 198–99
Avenida Paulista, 36
Avritzer, Leonardo, 78–79
Azevedo, Aluizio de, 109
Azevedo, Antonio Vicente de, 27
Azevedo, Francisco Ramos de, 142
Azevedo Marques, José Manuel de, 21–22,
    23–24, 26

Bacha, Edmar, 219
Backdi, 150
On Balance (Phillips), 167
Bannister, Jon, 200
Bar and Restaurant Union, 89–90
Bar de Esquina, 121–26, 129, 145, 155–57
Baring, João, 65–66
bar noise
  complaints about, 40–41
  fines, 116–17, 118–19, 126, 151–52, 161
  justifying, 9
  ordinances, 40–41, 99
  soundproofing, 90, 99
bar noise levels, measuring, 123–24, 126
Barreiros, Renato, 150–51, 180
Barros, Aldhemar Pereira de, 30–31
Barry, Peter, 61
bars
  defined, 90–91, 156–57
  inspections, 121–26
  leisure noise, 2–3
  1:00 a.m. Law for closing, 89–93, 95–96,
      99, 108–9, 116–17, 118–19, 123, 139–40,
      156–57, 212–13, 214
  Public Civil Action, 151–52
bass music, 173, 201
behavior, anti-social, 128
Békésy, Georg G., 47
Belindia, 219
Bell, Alexander Graham, 49
Bell Telephone Laboratories, 49–50
Bijsterveld, Karin, 3, 7
billboards, 120
Bio G3, 150
biopolitics, 210
The Body Multiple (Mol), 2

"Bonde da Juju" ("The Juju Crew), 150, 156–57
Bonduki, Nabil, 27–28
Boulos, Guilherme, 75
Bratton, William, 91–92, 190
Brazilian Acoustics Institute (*Instituto Brasileiro de Acústica*, IBA), 29, 30, 37–38, 41–42, 45
Brazilian Acoustic Society (Sociedade Brasileira de Acústica, SOBRAC), 58
Brazilian Communist Party, 78–79
Brazilian Institute of the Environment and Renewable Natural Resources (IBAMA), 57–58, 59
Brazilian National Standards Association (*Associação Brasileira de Normas Ténicas*, ABNT)
 creation of, 40, 56–57
 function, 55, 56–57
 nationalizing the, 56–57
 present day, 57
 Secovi and the, 61
 technical standards published, 39–40
Brazilian National Standards Association (*Associação Brasileira de Normas Ténicas*, ABNT) Civil Construction Committee. *See also* Acoustics Commission
 commissions, 58
 NBRs move, opposition to, 61–62
 purpose, 56–57
 standards, consensus approach to, 72
 technical standards, 40, 206, 211
Brazilian Regulatory Standards (*Normas Brasileiras Regulamentadoras*, NBRs), 55, 56–57, 59
Brazilian Social Democracy Party (PSDB), 136–37, 138–39
Brazilian Traffic Code, 191–92, 201
Brazil Is Not ForAmateurs (Castor), 130
Bresser Pereira, Luis Carlos, 130–31
broken windows theory, 14–15, 112, 169, 190
Burgess, C. F., 51–52

Caldeira, Teresa, 24, 111–12, 167, 168–69, 190
Callon, Michel, 12–13, 72
Camilo, Álvaro, 139–40, 190, 199
Campbell, Luther, 173
Campo Limpo subprefecture, 184, 186–88, 191

Canclini, Nestor Garcia, 157
Cano, Wilson, 20, 28
car audio, 176–80, 182, 200–1
Cardoso, Celso, 84
Cardoso, Fernando Henrique, 130–31
Cardoso, Henrique Fernando, 80–81
carjackings, 183–84, 185–86
Carneiro, Waldir de Arruda Miranda, 145–47, 148–49, 150–51, 155, 157–58, 160, 165–66, 208, 214
Carvalho, Deufrânio Barbosa de, 191, 192, 193, 194–95
Carvalho, Ricardo Torres de, 153–55, 164
Castelani, Debora, 116–17
Castor, Belmiro, 130, 135
Catholic Church, 79
Cavalieri Filho, Sérgio, 159
C. F. Burgess Laboratories, 51–52
chains of reference, 53
Chesluk, Benjamin, 112–13
Child and Adolescent Statute (*Estatudo da Criança e do Adolescente*, ECA), 193
Childress, Herb, 170
Chladni, Ernst, 48
church noise, 40–41, 95–96. *See also* religious noise
Church of the Grace of God, 79
Cidade Tiradentes, 174, 182–83
citizenship
 consumerism and, 28, 180–81
 differentiated, 7–8, 79, 215
 Evangelical organizations and, 79
 insurgent, 8, 79, 215–17
 noise and, 6–7, 98
 pancadão controversy and, 200
 urban, post-dictatorship Brazil, 79
 urban development and, 78–79
Civil Code, 147–48, 157–58
Civil Guard, 108–10
civil law
 agents interpreting, 146–47
 codes of, 147
 common law vs., 147, 148
 criminal law vs., 159
 proceedings, 148
 process, case study, 148–50
 right of appeal, 148
civil litigation, 80, 150

civil police (*Polícia Civil*)
  police report investigations, 152
  political dissidence, neutralizing, 110–11
  responsibility of, 104–5
Civil Procedure Code, 147–48
Clean City Law, 120, 134–35
clientelism, 129, 134–35, 136
Coelho, Bento, 94
coffee industry, 20, 93
collectivism, 9
Columbié, Orlando Vistel, 181–82
Commercial Code, 147–48
commercial venues
  licensing, 83
  noise fines, 83, 115–16
  noise ordinances, 83
common law, 146, 147
community noise
  assessing the acceptability of, 55
  complaints, 106
  fines, 151–52
  legislating, 3
  litigation, 146
  Public Civil Action, 151–52
  quermesses, 175–76
community policing, 112–13
Community Security Council (*Conselho Comunitário de Segurança*, CONSEG) meetings
  administrative flows, rewiring, 192–98
  Campo Limpo subprefecture, 186–88, 189
  Cidade Tiradentes, 182–83
  complainants, 185–86
  complaints about, 185–86
  Heliópolis, 186–88, 189
  Ipiranga subprefecture, 184–85
  Jabaquara subprefecture-district, 183–84, 189
  M'Boi Mirim subprefecture, 185–86
  minutes consistency, 182–83
  noise complaints heard, 185–91
  participants, 186, 189–90
Community Security Council (*Conselho Comunitário de Segurança*, CONSEGs), 112, 182–91
Congonhas Airport, 34–36, 42–44, 81, 205, 211
construction noise, 38–40, 164
construction sector

Acoustics Commission participation, 61, 69, 73
Acoustics Commission revised norms, 61
litigation, circumventing, 36–37
power of the, 69, 73
sanitation and hygiene concerns, 25
soundproofing link, 206
specifications for, 146
construction sonic complex, 206
consumerism, 28, 174, 180–81
consumption, conspicuous, 174, 180–81
corporatism, 130
Corten, André, 88
Corti, Alfonso, 47
Costa, Frederico Lustosa da, 130
Court of Justice chambers, 153
crime
  alcohol consumption and, 91–92, 214
  automobile, 183–84, 185–86
  broken windows theory of, 14–15, 112, 169, 190
  Campo Limpo subprefecture, 184
  fear of, 111, 190, 211–12
  funkeiros and, 181–82
  funk ostentação and, 182
  Jabaquara subprefecture-district, 183–84
  minority youth and, 199–200
  nightlife noise and, 207–8
  noise and, 118–19
  noise pollution as a, 80–81
  organized, 185
  peace disturbance and, 108–9
  reducing, 91–92
  scientific approach to, 106, 108–9
  spatializing, 112–13
  talk of, 190
  urban disorder and, 91–92
crime-culture, 212
crime maps, 105
Criminal Code, 147–48
criminal law, 152, 159
Criminal Law section, Supreme Judiciary Council (*Conselho Superior da Magistratura*), 142–44
Criminal Procedure Code, 147–48
cultural grooves, 171

Dadson, R. S., 50–51
Dagnino, Evelina, 8
DaMatta, Roberto, 7

Dantas, Konrad (Kondizilla), 154, 174
David, Captain, 112–13, 183–84
decibel
  authority of the, 214
  measuring, 30–31
  noise as, 128
  reference value, 50–51
  sound power as, 53–54
  use of, historically, 26
decibel values
  acceptable values for building and room types, 69
  standards controversy, 66–69
Del Carlo, Ualfrido, 39–40
Delegated Operation (*Operação Delegada*), 120, 139–40
democratic paradox, 3, 7, 9
democratization, challenging the limits of, 180–81
Department of Food Supply, 116–17
Department of Subprefectures, 116–17, 120, 139–40
Department of Urbanism, 32
Despretz, César-Mansuète, 47–48
Diffuse Interests Fund, 150–51
Diniz, Maria Helena, 158
disability-adjusted life-year (DALY), 53
discrimination, racial, 209, 212, 217
disorder, culture of, 190–91
Dória, João, 96
Douglas, Mary, 3–4
drug use, 175–76, 182

ear, human
  average, 12, 45–46, 128, 171–72, 210–11
  loud music, response to, 171–72
  passionate, 172
  primitive, 171–72
  replacing the, 54
  sculpture, 75, 99
  study of the, 47
Ear 1.0
  construction of, 47–48, 50–51
  defined, 45–46
  emulating, 53–54
  rest and health delinated by, 158
Ear 2.0. *See also* sound level meter (SLM)
  construction of, 51–52
  defined, 47–48

Ears 1.0 and 2.0
  circumventing, 201
  fine-tuning, 165
  inauguration of, 49–50
  introduction, 12
  loudness, exploring, 172
  relationship, 55
Echo Chamber. *See* São Paulo Municipal Chamber
Ecko, Marc, 174
economics
  anti-noise waves, 20–21, 211
  of bar closures, 89–90
  of health, 44
  noise, justifying, 9
Ed Hardy clothing, 180–81
Edison, Thomas, 49
entrepreneurial model, 133
Environmental Agency of the State of São Paulo (*Companhia Ambiental do Estado de São Paulo*, CETESB), 37, 59, 60, 116, 151, 206–7
Environmental Chamber, 153–55
environmental crime, 164–65, 192
Environmental Crime Law, 80–81, 164, 201
environmental noise
  converting into encoded principles, 144–45
  cost of, construction sector and, 206
  health, effects on, 44, 45–46, 53, 75
  individualism vs. collectivism and, 9
  informing notions of civility, 6–7
  legislation, 80–81
  management of urban, 6–7
  measuring, 52, 124
  measuring, technical standards of, 12
  misdemeanors, 162–63
  music events, 160–61
  Public Civil Action, 160–62
  punishment for, 164, 192
  traffic noise as, 94
environmental noise litigation, legal channels related to
  multi-channel sound politics, 189–90
  PSIU, fighting the, 145
  Public Civil Action, 145
  public law, 145
environmental noise litigation, legal channels related to agents of
  the appellate judge, 152–55

environmental noise litigation, legal channels related to agents of (*cont.*)
  the lawyer, 146–50
  legal scholars (*doutrinadores*), 146–47
  the state prosecutor, 150–52
environmental noise litigation, legal channels related to jurisprudence
  assessment of evidence, 158–59
  chambers reserved, 142–44
  environmental crime, 164–65
  misdemeanor law, 162–63
  private law, 157–60
  PSIU, fighting the, 155–57
  Public Civil Action, 160–62
environmental pollution, 34–35
equal-loudness contours, 50–52, 200–1
Erundina, Luiza, 81, 82
Ett, Steve, 173
Eucatex, 29, 30, 35, 39–40, 41–42, 206
Europe, noise pollution in, 6–7
Eustachio, Bartomeo, 47
"Evaluation of Noise in Inhabited Areas Aiming for the Comfort of the Community, 71
Evangelical churches
  citizenship and, 79
  noise, justifying, 9
  noise fines, amnesty for, 96
  noise laws, 96
  power of, 80–81, 85–86, 98
  sonic complex, 207–8
  sonic excess and salvation, 88–89

Fabricius, Hieronymus, 47
favelas, 181–82
Federal Housing and Urban Planning Service (*Serviço Federal de Habitação*), 34
Feld, Steven, 171
Fellini, 16
Feltran, Gabriel, 79
Fernandes, Heloisa, 111
Financing Fund for Integrated Local Development Plans (*Fundo de Financiamento de Planos de Desenvolvimento Local Integrado*), 34
"First Municipal Conference on Noise, Vibration, and Sound Disruption," 75–77

Fletcher, Harvey, 50–51
Fletcher-Munson curves, 50–51, 52
Folha de São Paulo, 19, 38
Fonseca, Luiz, 23–24
Fortaleza, 94
Foucault, Michel, 14, 102–3, 128
Fowler, Edmund, 50–51
Free, Edward E., 51–52
Freire, Ms., 122–23, 124–26, 127–28, 129, 155–56, 183–84, 217–18
Freyre, Gilberto, 209
Friends of the City Society (*Sociedade Amigos da Cidade*, SAC), 26, 41–42
Funk Brasil Vol. 1 (DJ Marlboro), 173–74
funk carioca, 14–15, 173–75
funk carioca youth, 170
Funk Festival, 180
funk ostentação
  bass in, 201
  beginnings, 173–75
  crime and, 182
  explosion, 180
  factors underlying, 177–78
  mobile *pancadão and*, 177
  social media in organizing, 217
  spread of the, 182
Funk Parties, 180, 192–93, 199

Gaebler, Ted, 131, 133
Gaetani, Francisco, 135
Galileo Galilei, 48
Galton, Francis, 47–48
General Planning Office, 37
General Supervision of Land Use and Occupation, 120
Gilroy, Paul, 179–80
Giuliani, Rudolph, 91–92, 190
glossolalia, 88
Gomart, Emilie, 170–71
González, Velda, 181–82
Goodman, Steve, 149–50
government, executive branch, sources of administrative instability
  budgetary concerns, 132–35
  legal-bureaucratic arrangements, 128–31
  party affiliation, 135–41
government, noise and the, 5–6, 44, 45–46, 80–81

graffiti, 167
Green Party, 82
Guardianship Council, 192–93
Guércio, Arthur del, 157

Haddad, Fernando, 77, 95–96
happiness, 89–90
Hartmann, Arthur, 47–48
Hato, Jooji, 89, 90–93, 94–96, 99, 116–17, 176, 177, 179–80, 198–99, 214–15
health
  economic rationale, 44
  noise, effects on, 44, 45–46, 53, 75, 81, 82–83, 164, 165, 214
hearing, 47–48
hearing, territoriality and, 3–4
hearing defects, 29
hearing loss, 75
hearing loss compensation, 53
hearing threshold, 53–54
Heliópolis, 184–85, 186–88, 189
Helmoltz, Hermann, 47, 48
Hennion, Antoine, 170–71
Henriques, Julian, 172
Heredia, Blanca, 135
hip-hop, 173, 177–78
Holloway, Thomas, 109–10
Holston, James, 7, 8, 111–12, 215
homeless people, 190
home-ownership model, 32, 34
homicide, rates of, 92, 183–84, 185
Hotels, Restaurants, and Bars Union, 90–91
Housing and Urban Development Company, 184–85
Housing Union (Secovi), 61
Houston hip-hop, 177–78
Hull, Matthew S., 11
hybrid forums, 72
Hyrtl, Joseph, 47

individualism, 7, 9
industrialization, 32, 59–60, 68–69
industrial noise, 68–69, 81, 207
industrial sonic complex, 206–7
industry, hygiene and safety regulations, 24
infrastructure, 93, 184–85, 219
Ingana, 219

An Inquiry into Modes of Existence (Latour), 15–16
inscription device, 46
Institute of Technological Research *(Instituto de Pesquisas Tecnológicas*, IPT), 35, 37–38, 39–40, 41–42, 45
International Electrotechnical Commission (IEC), 51–52, 56
International Noise Awareness Day, 81–82
International Organization for Standardization (ISO), 56
Ipiranga subprefecture, 184–85
Ising, Hartmut, 3
Isnard, Nicolas, 94

Jabaquara subprefecture-district, 181–82, 183–84, 188, 189
Jardins districts, 24, 28–29, 36
John the Baptist, celebrating, 175–76
judiciary
  the appellate judge, 152–55
  assessment of evidence, 158–59
  civil law, 148
  command center, 142–44
  constitutional autonomy, 159
  dilemmas in, 213
  police report investigations, role in, 152
  restaurant noise complaints, litigating, 149–50
  role of the, 148
Juliet sunglasses, 174

Kang, Jian, 3
Kassab, Gilberto, 87–88, 139–40
Keil, Charles, 171
Kelling, George, 14–15, 112, 169
Koenig, Rudolph, 48, 49
Kruppa, Barbara, 3

LaBelle, Brendon, 179–80
Land Utilization Index (LUI), 36
Laplace, Pierre-Simon, 48
Larkin, Brian, 2–3
Latour, Bruno, 9–10, 15–16, 46, 53, 54, 96–97, 127, 144, 204, 215, 218–19
law
  average body, defining, 128
  corruption of the, 134–35
  tangential approach to, 144

lawmaking process, 134–35
lawyers
  civil law, 148
  environmental, 146–50
Lefebvre, Henri, 8
legislation. *See also* noise legislation
  anti-*pancadão*, 15
  enforcing, 31
  environmental noise, 80–81
  pancadão, 181–82
  political pragmatism in, 31–34
legislation, urban
  elite vs. suburban/rural zones, 24, 28–29
  spatial segregation, 24, 28–29
  "Urban Noises, the Location and Operation of Obnoxious, Harmful, or Dangerous Industries" Bill 335, 30
leisure noise, 2–3, 89–90. *See also* bar noise
Lenke, Thomas, 14
Lima, Vicente Faria, 31–32
listeners, average, 50–51
listening-in-readiness, 211
listening-in-search, 211
Lopes, Conte, 199
"Loud Car Stereos" (DOJ), 200–1
loudspeakers, 49
Lutti, José Eduardo Ismael, 127, 135, 150, 151, 152, 155

Macedo, Regina, 81
Mainwaring, Scott, 136, 140–41
malls, *rolezinhos* in the, 217
Maluf, Paulo, 35, 83–84, 85, 90, 116, 206
Marc Ecko watches, 174
Marques, Azevedo, 41–42
Massumi, Brian, 101–2
Master Plan
  Advising Commission, 32
  constitutional requirement, 78–79
  Land Utilization Index (LUI), 36
  legislation, politics in, 10
  litigation circumventions, 36–37
  noise component, 37, 41
  noise maps, 13, 94–95
  residential input, 36–37
  revisions, support for, 77–78
  zoning laws, 36–37
Matarazzo, Andrea, 75–77, 93, 94–95

material compensation, 159
Matless, David, 112–13
Mattos, Juvenal Lino de, 30
M'Boi Mirim subprefecture, 185–86, 191, 192–98
"Measurement and Evaluation of Sound Pressure Levels in Indoor Environments," 71
"Measurement and Evaluation of Sound Pressure Levels in Inhabited Areas," 71
"Measurement and Evaluation of Sound Pressure Levels in Outdoor Environments," 71
Meirelles, Hely Lopes, 161
Mendes, Beto, 185
Merryman, John Henry, 148
Metropolitan Civil Guard, 110
Miami bass, 173–74
Micro-Environmental Technology Lab, 35
microphone, 49
middle class, 180–81, 209
Miles of Standard Cable (MSC) unit, 49–50
military police (*Polícia Militar*)
  budget, 132, 139
  civil actions against, 151–52
  coercive methods, 102, 108–11
  community policy strategies, 112–13
  community respect for, 107
  corruption, 127–28, 134–35
  crime maps, 105
  Delegated Operation (*Operação Delegada*), 139–40
  disciplinary infrastructure, 128–29
  historically, 108–11
  inspections, off-duty, 120
  introduction, 13–14
  legal obstacles, circumventing, 112–13
  legal stability to punish, 128, 129
  loud music, actions to repress, 15
  noise, responsibility for, 27
  operations, types of, 105
  pancadãoes and, 181–82, 189–98
  products and services, 133
  PSIU and the, 14, 118–19, 139–41
  PSIU vs., 102
  public opinion, manipulating, 112–13
  purpose, 110–11
  resource allocation, 105, 107, 128

responsibility of, 104–5
scientific approach to crime, 106, 108–9
security council meetings, 112–13
statistics, authority of, 106, 128–29
structure, 105
torture methods, 110–11
training, 112
transparency, 128–29
violence by, 108–12
zero tolerance policy, 91–92, 119, 137, 138–39, 169
military police (*Polícia Militar*) noise complaints
administrative flows, clogging of, 106, 107
complaint procedure, 103–4
involvement in, 186–88
police actions, 104, 106
reports, 106
statistics, 106
Military Police Operations Center (*Centro de Operações da Polícia Militar*, COPOM)
administrative flows, clogging of, 106, 107
complaint procedure, 103–4
decision to engage, 104–5
online complaint system, 107
surveillance cameras, 128
Miller, Peter, 14
Mills, Mara, 50–51
Minas Gerais, 57–58
minhocão expressway, 35–36, 42–43, 81, 205
minority age debate, 199–200
Miraglia, Paula, 92
Misdemeanor Law, 162
misdemeanor legal channels, 162–63
modernity
construction projects in urban, 206
differentiated citizenship and, 215
inconsistencies in São Paulo, 219
Latour's approach to, 16
recalling and resetting, importance of, 219
modern life, sounds in, 2–3
Mol, Annemarie, 2
moral damage compensation, 150, 151–52, 159, 161–62
Moreno, Gilberto, 176
motoboys, 167–68
motorcycle police force (*Ronda Ostensiva com Apoio de Motocicletas*, ROCAM), 105

Mouffe, Chantal, 9–10
Mr. Mixx, 173
Municipal Housing Agency, 184–85
Munson, Wilden A., 50–51
music, anticipating socioeconomic developments, 4
music, loud. *See also* funk carioca, funk ostentação; funk ostentação; *pancadões*
complaints about, 41
ear's response to, 172
human body boundaries and, 172
introduction, 14–15
police actions to repress, 15
sound-politics, place in, 170
sub-bass, 173, 177–78
musical grooves, 170–73, 182
music events, 160–61
music videos, 174

Nascimento, Gilberto, 84
National Council of Metrology, Standardization, and Industrial Quality (*Nacional de Metrologia Normalização e Qualidade Industrial*, Conmetro), 57
National Council of the Environment (CONAMA), 57, 59
National Housing Bank (*Banco Nacional de Habitação*), 34
National Institute of Metrology Standardization and Industrial Quality (*Instituto Nacional de Metrologia, Qualidad e Tecnologia*, Inmetro), 57–58
National Movement for Urban Reform, 78–79
National System of Metrology, Standardization, and Industrial Quality (*Sistema Nacional de Metrologia, Normalização e Qualidade Industrial*, Sinmetro), 57
National System of the Environment (SISNAMA), 57
National Traffic Council (CONTRAN), 191–92, 201
neighborhood, defining, 158
neighborhood law (*direito de vizinhança*), 80, 145, 157–58

neighborhood noise, 81–82, 146
neighborhood-unity organization, 32–34
Neto, Antônio Delfin, 219
New Acoustics, 46–50, 178
New Public Management, 130–31
New Republic period, 130
New York City, 22, 32, 36–37, 91–92, 190
New York City Noise Abatement Commission, 51–52
New York City Noise Code, 81–82
NGOs, 78–79
nightlife noise
  debate over, 89–93, 99
  legislating, 207–8, 211
  religious noise vs., 92–93
nightlife sonic complex, 207–8, 212–13
Nike Shox, 174
Nogueira, Ataliba, 163
noise
  behavioral, 29–31, 40–41
  citizenship and, 98
  crime and, 118–19
  defining, 208–9
  defining, attributes for, 3
  factory, categories of, 30
  from within buildings, 65
  gestures of, 204
  government involvement in, 5–6, 44, 45–46, 80–81
  health, effects on, 12, 29–30, 75, 164, 165, 214
  heterogeneity of, 7, 10–11
  high- vs. low-frequency, 200–1
  infrastructural, 29–30, 31, 35, 41
  justifying, 9
  negotiating, 204
  ontological multiplicity of, 2, 3–4, 14
  pancadão, 185–86
  parasitic nature of, 5–6, 10–11, 204
  in the press, 19, 21, 26, 40–41
  punishment for, lack of, 21
  quantifying, 29–30, 31
  sound-as-, 3
  term, removing from standards, 71
  violence and, 77
noise abatement, 3, 6–7
noise-as-decibel, 128
noise-as-incivility, 128–29
noise-as-nuisance, 128

noise-as-waste, 3
noise circles, 96–100
noise complainants
  complaint procedure, 103–4, 106, 107
  fears of retaliation, 80, 107
  identifying, 84
  material compensation, 159
  moral damage compensation, 150, 151–52, 159, 161–62
  reporting process, 104–5
noise complaints
  administrative entities responsible for, 13–14
  against churches, 84–86
  centralizing, 81
  CONSEG meetings, 185–91
  defense strategies, 162–63
  enforcement, 80
  legally valid, 158
  litigation process, 148–50
  military police involvement in, 104, 106, 118–19
  pancadão, 175, 176, 182–83
  police involvement in, 186–88
  police reports about, 106, 107, 162
  Public Civil Action, 151
  restaurants, litigating, 148–50
noise complaints, appealing
  Bar de Esquina, 155–57
  provisional and final sentences, 152–53, 154
  PSIU, fighting the, 155–57
noise control, passive vs. active approach to, 64
noise controversies
  administrative response to calls, 11–12
  calls from residents/groups to eliminate sonic practices, 11–12
  entities heard, potential, 218
  governmental solutions, 213–14
noise-crime axis of debate, 211–12
noise-culture axis of debate, 208–10, 212
noise damage, measuring, 53
noise debates
  focus of, 78
  nightlife noise, 89–93, 99
  noise laws, 78–84
  religious noise, 84–89, 98
  traffic noise, 93–96, 99
noise exposure limits, 69

Index 241

noise fines
  abusive, 132
  amnesty for, 96
  appealing, 126, 145, 161
  bars, 99, 116–17, 126, 151–52
  car music while parked, 198–99
  car noise, 191–92
  church buildings, 85–87
  commercial venues, 83, 115–16
  community noise, 151–52
  executing, time limits on, 157
  inspections and, 81
  limits on, 38
  misdemeanors, 162
  music from parked cars, 198–99
  religious, amnesty for, 96
  religious noise, 87
  restaurants, 149
  statistics, 38, 132
noise-health axis of debate, 210–11, 212
noise infractions, punishment for, 162, 163–64
Noise Law, 115–16, 117–18, 123, 212–13
noise laws, debate over, 78
noise legislation. *See also* 1:00 a.m. Law
  enforcement, 80, 81
  noise limits, 30–31, 37–38
  pancadãoes, 199
  paradox of control in, 3
  ProAcústica lobby for, 41
"Noise Levels for Acoustic Comfort (NBR), 55, 71
noise limits
  decibel specifics, 37–38
  industrialization and, 68–69
  paradox of control in, 7
  rural areas, 68–69
  standards, 68–69
  timetables, 68–69
noise limits legislation, 30–31, 37–38
noise litigation, 14, 213
Noise Map Law, 95
noise maps
  active approach to noise control, 64, 99
  developing, 60
  European cities, 6–7
  Master Plan, 13
  resistance to, 95
  support for, 94–95

noise-market axis of debate, 211, 212
noise measurement
  elements of, 3
  from within buildings, 65
  in litigation process, 149
  noise-health axis of debate, 210–11
  outside vs. inside, 70
  pancadão, 169–70
  reference value for, 65–66
  requirements for, 162–63
  SLM (sound level meters), 123
  sound of churches, 85–86, 87–88
  standards, 65, 210–11
  technology, 51–55, 60, 71–72, 200–1
  with accountability, 53
  zoning laws and, 87–88
noise measurement standards, 55–58
  controversy, 69–70
  vocabulary changes, 70–72
noise ordinances
  agencies designing, 37
  behavioral traffic noise, 23–24
  churches, 40–41, 84–89
  commercial venues, 83
  enforcing, 84, 92
  hours of operation, 24
  industry, 24
  religious noise, 40–41, 84–89
  revising the, 81–84
  soundproofing, 85–86
  zoning laws and, 37–38
noise persistence, premises for
  governmental lenience, 42
  infrastructure deficiency, 42–43
  introduction, 41–42
  lack of human civility, 42
  lack of law-enforcement, 42
  technological obsolescence, 43–44
noise pollution
  anti-noise waves, 34–35
  conference on, 75–78
  Congonhas Airport, 35–36
  criminal, 80–81
  European, 6–7
  health, effects on, 81, 82–83
  jurisprudence, 165
  legislation, 80–81
  minhocão expressway, 35–36
  mitigating, 94

Noise: The Political Economy of Music (Attali), 4
noise typology, 26
Novak, David, 172
nuisance, 27, 128

objects, ergonomopolitics of, 50–51
O Estado de São Paulo, 19, 40–41
1:00 a.m. Law, 89–93, 95–96, 99, 108–9, 116–17, 118–19, 123, 139–40, 156–57, 212–13, 214
ontologies, term usage, 2
Oosterbaan, Martijin, 88–89
Orchestra Rehersal (Fellini), 16
Osborne, David, 131, 133

Palace of Justice (*Pálacio da Justica*), 142–44
pancadão variants
  mini-*pancadão*, 175
  mobile *pancadão*, 176–77
  staged *pancadão*, 175–76
pancadão vehicles, 176–80, 182, 191–92
pancadões
  administrative flows, rewiring, 192–98
  anti-*pancadão* operations, 191–98
  articulating the, 175–81
  CONSEG meetings, 183–86, 188
  crime and, 212
  culture of disorder, 190–91
  disarticulating, groups involved in, 15
  disarticulating the, 169, 181–82, 192–93
  ears and norms, rewiring, 200–1
  echo chamber, rewiring the, 198–200
  family values and, 189–91
  funk carioca and funk ostentação, 173–75
  introduction, 14–15
  legal channels, rewiring, 191–92
  legislating to extinguish, 199
  legislation, 181–82
  loud grooves, 170, 172–73
  measuring the, 169–70
  migration to the, 181–82
  police and, 15, 181–82, 189–98
  popularity of, 168–69
  race-class-space continuum and, 186
  in slums, 188
  social media in organizing, 176–77, 217
  sonic dominance as groove attachment, 171
  zero tolerance policy, 200

pancadões controversy
  family values and the, 189, 210
  funk carioca and funk ostentação, 173–75
  loud grooves, 170, 172–73
  popularity of, 168–69
  race-class-space continuum and the, 186
Parables for the Virtual (Massumi), 101
passion, 170
peace disturbance
  allocation of resources for, 107–8, 119
  crime and, 108–9
  defined, 103, 104
  from vehicles, 201
  jurisprudence, 165
  legal category, 104
  statistics, 106
Penal Code, 147–48
Penal Contravention Laws (PCL), 103, 104
Pentecostalism, 88
Pereira, Alexandre, 175
Pereira, Bresser, 130–31
Pereira, Wanderley, 119
Pérez-Pérdomo, Rogelio, 148
Permanent Police Corps (*Corpo Policial Permanente*), 109–10
perspectivism, 2
Perturbações Sonoras nas Edificações Urbanas ("Sonic Disturbances in Urban Buildings") (Carneiro), 145–46
Peterson, Marina, 34–35
Phillips, Adam, 167
phonoauthograph, 49
phonograph, 49
pimp persona, 177–78
Pitta, Celso, 90–91
pixação, 167
pixadores, 167
PMDB (Party of the Brazilian Democratic Movement, 138
Police Real Guard, Military Division, 109
Policing Planning (PPI), 105
political, term usage, 9–10
political prisoners, 110–11
politics
  defined, 96–97
  meaning of, 9–11
Pollution Control Supervision Board, 37

post-structuralism, 8
poverty, 180–81, 212. *See also* slums
press, noise in the, 19, 21, 26, 40–41
private law, 157–60
Private Law section, Supreme Judiciary Council (*Conselho Superior da Magistratura*), 142–44
privatization, 130–31
ProAcústica, 13, 41–42, 45, 59–60, 94
property, misuse of, 158, 160–61
property rights, 44, 80, 157–58
public advertising, 120
public buildings, soundproofing, 81
Public Civil Action
   bars, 151–52
   community noise, 151–52
   environmental noise, 160–62
   environmental noise litigation, 145
   noise complaints, 151
Public Civil Action Law, 150–51, 159
public ear, 12
Public Force (*Força Pública*), 110
Public Law Chamber, 153–54, 155
Public Law section, Supreme Judiciary Council (*Conselho Superior da Magistratura*), 142–44
public nuisance link to zoning, 27
public order, offenses against, 109–10
Public Orientation Service (*Serviço de Orientação ao Público*, SOP), 104
public-private partnerships, 130
Public Prosecutor's Office (*Ministério Público*, MP), 145, 150–52, 160–61
public security
   Civil Guard for protecting, 110
   Delegated Operation for, 139
   nightlife noise complex and, 207–8
   1:00 a.m. Law and, 92
   pancadãoes, disarticulating for, 169, 190–91, 199–200
   police coordination for, 104–5, 106
   police violence justifying, 111
   post-redemocratization, 111
   right to, 189
   sound-politics and, 113
   spending, 132
   street sonic complex and, 208
   zero tolerance policy, 91–92
public security debates, 207–8
Public Security Secretary, 106, 111, 132, 182–83
public servants, career regimes, 135
public space, 30, 199–200
   auditory practices in, 2–3
Pythagoras, 48

Quadros, Jânio, 30, 31–32, 110
Queiroz, Tecyo, 178
quermesses, 175–76
quiet zones, 94–95

racial mixing, 209
radio, 22
rap music, 173–74
Rayleigh, Baron, 48
Reassembling the Social (Latour), 204
reggae, 173
Regional Administrations era, 138
Reinventing Government (Osborne & Gaebler), 131
relativism, 2
religious freedom, 79, 80–81, 84
religious noise
   complaint and response procedure, 113–15
   complaints about, 40–41
   debate over, 84–89
   fines, amnesty for, 96
   glossolalia, 88
   justifying, 9
   legislating, 207
   limits on, 95–96
   negotiating, 134–35
   nightlife noise vs., 92–93
   noise debates, 98
   ordinances, 40–41, 84–89
religious sonic complex, 207
residential noise complex, 208
Respect Action Plan, 200
restaurant noise complaints, litigating, 148–50
re-urbanization, 184–85
Robinson, D. W., 50–51
Robinson-Dadson curves, 50–52
rolezinhos, 217
Rolnik, Raquel, 25
Roma, Celso, 138

244 Index

Roofless Workers' Movement (*Movimento dos Trabalhadores sem Teto*, MTST), 77, 95, 98
Rosado, Moacir, 117–19, 129
Rose, Nikolas, 14
Rose, Tricia, 173
Rosemberg, André, 109–10
Rosenblith, Walter A., 47

Sabine, Wallace, 48
Salim, Dito, 84
sanitary movement, 25
São Paulo
  growth and urban density, 18, 20
  regional administrations, 36
  urban growth, 27–28
São Paulo Anti-Noise Agency (*Programa de Silêncio Urbano*, PSIU)
  administrative fines, 145
  administrative flows, 121–26, 127, 155–56
  anti-*pancadão* operation, 192–93, 194–95, 196
  budget, 132
  bureaucracy, 120–22
  career regimes, 135
  civil actions against, 151–52
  complaint and response procedure, 113–15, 120
  complaints about, 77
  complaint statistics, 119
  corruption, 127–28, 134–35
  creation of the, 83–84, 116
  criticisms of the, 77, 150, 151
  decentralization/recentralization, 116–18
  disciplinary infrastructure, 129
  enforcement function, 217–18
  expansion of the, 117–18, 119
  fighting through legal channels, 145, 155–57
  fine registration sector, 120–21
  function of, 102
  institutional inconstancy, 129
  introduction, 13–14
  juridical sector, 120–21
  legal stability to punish, 129
  meeting demands, 127
  noise maps, support for, 60
  1:00 a.m. Law and, 116–17
  passive approach to noise control, 64
  police and the, 14, 118–19, 139–41
  police vs., 102
  political interference, 135–36
  programmers, 121–22
  religious noise complaints, 113–15, 207
  responsibilities, 120
  structure, 116
  subprefecture system, impact of, 116–17, 120, 132, 137–38
São Paulo Building Code, 24
São Paulo Municipal Chamber
  lawmaking process, 13
  noise debates, 13
  noise-related bills, 13
  pancadãoes rewiring, 198–200
São Paulo Municipal Chamber Bullet Caucus (*Bancada da Bala*), 199–200
Scarpa, Antonio, 47
Schafer, Murray R., 2–3, 4
schizophonia, 2–3
science
  average body, defining, 128
  knowledge and, 15–16
  limitations of, 72
  modern, resolving social issues, 43–44
scientific management, 27–28
Scott, Édouard-Léon, 49
Scott, Michael S., 200–1
On Sensations of Tone (Helmoltz), 48
Serres, Michel, 5–6
Siciliano, Heribaldo, 26–27, 41–42
Siemens, Ernst W., 49
signal loss, 49–50
Silence Commission, 26
Silent Week campaign, 29
Sinduscon, 61
sleep disturbances, 6–7
slums
  Campo Limpo subprefecture, 183–84, 191
  emergence of, 27–28
  Ipiranga subprefecture, 184–85
  Jabaquara subprefecture-district, 181–82, 183–84
  M'Boi Mirim subprefecture, 185, 191
  pancadãoes in, 188
Sobral, Helena, 82
Sobral, Regina Macedo, 116
social media, 176–77, 217
social rights, 7
sonic behavior, study of, 48

sonic classifications, 2–3
sonic complexes
　industrial, 206–7
　nightlife, 207–8, 212–13
　religious, 207
　residential, 208
　street, 207–8
　traffic, 205–6
sonic events, subjective vs. objective, 3
sound
　filtering, 211
　measuring, protocol for, 55
　over distance, 49
　pain threshold, 50
　in sound-politics, 4, 204
sound-as-noise, 3
sound level meter (SLM), 12, 46, 51–55
　A and C weightings, 200–1
　providers of, 60
　vocabulary changes, 71–72
sound media, 48–49
sound-politics
　defined, 215
　instituting, 18–19, 26
　lawmaking in, 97
　modes of existence, 204
　multi-channel, 165–66
　of music, 170, 171
　music, place in, 170
　public security and, 113
　sound in, 4, 204
　statistical approach to, 129
　term usage, 1–2
　underlying foundation, 10
sound-politics, four strata of
　axes of debate
　　noise-crime, 211–12
　　noise-culture, 208–10, 212
　　noise-health, 210–11, 212
　　noise-market, 211, 212
　governmental dilemmas, 212–13
　governmental solutions, 213–15
　introduction, 204–5
　recommendations for studying, 218–19
　sonic complexes, 205–8
sound pressure levels, establishing, 30–31
soundproofing
　anti-noise waves, 29, 30, 60
　bars, 90, 99
　church buildings, 85–86
　requirements, 81
soundproofing market, 39–40, 41, 60, 206
sound reproduction technologies, 46–47
soundscape, modern, 12
The Soundscape of Modernity (Thompson), 12
sound standards, 55–58
sound studies, 4–5
sound technology, misuse of, 2–3
South-West axis, 32–34
space, criminalizing, 112–13
spatial segregation, 24, 28–29, 32–34, 77–78, 179–80, 184
speech, commercially acceptable, 49–50
state prosecutor (*promotor de justiça*), environmental noise litigation, 150–52
Stengers, Isabelle, 218–19
Sterne, Jonathan, 4–5, 48
street sonic complex, 207–8
Suplicy, Marta, and administration, 85–86, 87, 92, 116–17, 132, 134–35, 138
Supreme Judiciary Council (*Conselho Superior da Magistratura*), 142–44

tagging, 167
technical standards
　ABNT, 40, 175–76
　acoustic comfort, 40
technology
　anti-noise, 6–7
　availability of, 31, 38
　sound reproduction technologies, 46–47
Tedoro, Ms., 121–22
teenagers, rowdy. See also *pancadão*
　introduction, 14–15
　motoboys, 167–68
　pixadores, 167
　recreating public space, practices for, 167–68
　summary overview, 201–3
telegraph, 49
telephone, 49–51
Telhada, Paulo, 199–200
Tenório de Almeida, Carlos, 106, 107–9, 113, 119, 128–29, 162
territoriality, hearing and, 3–4
Thompson, Emily, 12
Todd, Neil, 171–72
torture, military police, 110–11

traffic accidents, 167–68
traffic complex, power of the, 38–39
Traffic Engineering Agency (*Companhia de Engenharia de Tráfego*, CET), 108–9
traffic infractions, 108–9
traffic noise
    behavioral, 22–24
    debate over, 93–96, 99
    elimination of, 25
    honking, 28–29
    increase in complaints against, 28
    infrastructural, 22
    legislating, 38
    loopholes, 65
    motoboys, 167–68
    noise-market axis of debate, 211
    public reporting on, 26
    punishment for, 191–92
    standards, 64–66
    state protection of, 206
traffic sector, 73
traffic sonic complex, 205–6
translation gaps, 19
Transmission Unit (TU), 49–50
Tripoli, Roberto, 81–84, 87, 88–89, 90, 92–93, 94–95, 98, 212–13, 214
Truax, Barry, 211
Tuma, Eduardo, 96
tuning fork, 47–48, 49

Universal Church of the Kingdom of God, 79
urban centers, auditory practices in, 2–3
Urban Guard (*Companhia de Urbanos*), 109–10
urbanization/urban development
    anti-noise waves and, 20
    car-based, 28
    citizenship and, 78–79
    construction sonic complex and, 206
    European, 6–7
    insurgent citizenship and, 215–17
    noise legislation, 30
    scientific management of, 27–28
    state-sponsored, 184–85
Urban Noise Control Superintendence, 38
urban planning, accountability in, 35
Urban Silence Program (*Programa de Silêncio Urbano*, PSIU). *See* São Paulo Anti-Noise Agency (*Programa de Silêncio Urbano*, PSIU)

urban space
    hygiene and safety regulations, 25
    regressive/immoral vs. progressive/moral behavior in, 21–22
    social production of, 8
    spatial segregation, 24

Valsalva, Antonio Maria, 47
Vargas, Getúlio, 27, 129–30
Vesalius, Andreas, 47
Vieira, Paulo, 28–29
Vila Madalena, 89–90
violence
    military police (*Polícia Militar*), 108–12
    nightlife and, 99
    noise and, 77
visual pollution, 120

Walser, Robert, 172
Weberian bureaucratic model, 129–31
Wegel Robert, 50–51
Weiner, Isaac, 88
welfare state, 130
Western Electric, 50–51
Wien, Max, 47–48
Willis, Thomas, 47
Wilson, James Q., 14–15, 112, 169
windows, noise-canceling, 60
Wolfe, Joel, 28
Wollaston, William Hyde, 47–48
Woolgar, Steve, 46
Workers' Party (*Partido dos Trabalhadores*, PT)
    activism, 78–79, 82
    creation of, 110–11
    Delegated Operation and the, 139–40
    emergence, 8–9, 136–37
    growth of the, 136–37
    ideology, 137
    noise maps, resistance to, 95
    religious noise and the, 85–86, 87
workplace noise, 69, 211
World Health Organization, 6–7, 53, 210

youth. *See also pancadaões*
    as disruption, 189–90, 199–200
    marginalizing, 170
    policymaking, incorporating in, 170

youth population
   motoboys, 167–68
   pixadores, 167
   recreating public space, practices for, 167–68
   summary overview, 201–3
YouTube music videos, 174
Yúdice, George, 182

01dB group, 60
zero tolerance policing, 91–92, 119, 137, 138–39, 169, 190, 212
zero tolerance policy, 200
zoning laws
   blueprint, 31
   changes to, 95–96
   creation of, 95–96
   debate over, 32–34
   industrial noise and, 81
   Master Plan, 36–37
   noise level limits, 87–88
   noise ordinances, 37–38
zoning-public nuisance link, 27, 208–9

www.ingramcontent.com/pod-product-compliance
Ingram Content Group UK Ltd.
Pitfield, Milton Keynes, MK11 3LW, UK
UKHW042006230426
12048UKWH00009B/583